SOUTH BOUND

11 [2] Daily Ex Sund'y	1 [3] Daily	11 [3] Daily Ex Sund'y
AM	AM	AM
5 53	6 00	6 13
......	......	......
5 55	6 03	6 15
5 56	......	s 6 16
5 57	......	f 6 17
5 58	......	s 6 18
5 59	......	6 19
.....	f 6 06	......
.....	f 6 08	......
.....	f 6 09	......
.....	f 6 12	......
.....	f 6 13	......
.....	f 6 15	......
.....	f 6 17	......
.....	f 6 20	......
.....	f 6 22	......
.....	s 6 25	......
.....	f 6 26	......
.....	f 6 29	......
.....	f 6 31	......
.....	s 6 33	......
.....	f 6 35	......
.....	s 6 38	......
.....	6 41	......
AM	AM	AM

TIME TABLE NO. 9 SOUTH BOUND—Continued

3 [1] Daily Ex Sund'y	11 [4] Daily	5 [4] Daily Ex Sund'y	31 [1] Daily Ex Sund'y	11 [5] Daily	7 [1] Daily	11 [6] Daily	21 [1] Daily Ex Sund'y	9 [2] Daily	11 [7] Daily
AM	AM	AM	AM	AM	AM	AM	AM	AM	AM
6 20	6 33	6 40	6 45	6 53	7 00	7 13	7 19	7 20	7 33
......	......	......	6 48	......	......	......	......	......	......
6 23	6 35	6 43	......	6 55	7 03	7 15	......	7 23	7 35
......	s 6 36	......	......	s 6 56	......	s 7 16	......	......	s 7 36
......	f 6 37	......	......	f 6 57	......	f 7 17	......	......	f 7 37
......	s 6 38	......	Via Tunnel	s 6 58	......	s 7 18	Via Tunnel	......	s 7 38
......	6 39	......		6 59	......	7 19		......	7 39
f 6 26	......	f 6 46	......	......	f 7 06	......	......	f 7 26	......
f 6 28	......	f 6 48	6 53	......	f 7 08	......	7 23	f 7 28	......
f 6 29	......	f 6 49	......	......	f 7 09	......	......	f 7 29	......
f 6 32	......	f 6 52	......	......	f 7 12	......	7 27	f 7 32	......
f 6 33	......	f 6 53	......	......	f 7 13	......	......	f 7 33	......
f 6 35	......	f 6 55	7 02	......	f 7 15	......	7 29	f 7 35	......
f 6 37	......	f 6 57	......	......	f 7 17	......	......	f 7 37	......
f 6 40	......	f 7 00	......	......	f 7 20	......	7 32	f 7 40	......
f 6 42	......	f 7 02	......	......	f 7 22	......	......	f 7 42	......
s 6 45	......	s 7 05	s 7 14	......	s 7 25	......	s 7 35	s 7 45	......
f 6 46	......	f 7 06	FREIGHT 2d Class Train	......	f 7 26	......	LIMITED 1st Class Train	f 7 46	......
f 6 49	......	f 7 09		......	f 7 29	......		f 7 49	......
f 6 51	......	f 7 11		......	f 7 31	......		f 7 51	......
s 6 53	......	s 7 13	7 25	......	s 7 33	......	7 41	s 7 53	......
f 6 55	......	f 7 15	......	......	f 7 35	......	......	f 7 55	......
s 6 58	......	s 7 18	7 31	......	s 7 38	......	s 7 46	s 7 58	......
7 01	......	7 21	7 34	......	7 41	......	7 49	8 01	......
AM	AM	AM	AM	AM	AM	AM	AM	AM	AM

LAUREL LINE
39
FREDERICK
BEAVER 80

LAUREL LINE

An Anthracite Region Railway

INTERURBANS SPECIAL 103

LAUREL LINE

An Anthracite Region Railway

by James N.J. Henwood and John G. Muncie

INTERURBAN PRESS • GLENDALE, CALIFORNIA • 1986

DUST JACKET AND FRONTISPIECE

Artist Frederick Beaver has successfully captured the mood of a northeastern Pennsylvania autumn afternoon in this rendering of the Laurel Line's newest steel passenger car as it passes through the woods near Moosic in 1952. The oil painting is based on a photograph by Edward S. Miller, the celebrated chronicler of North America's best known third-rail interurban.

FRONT ENDSHEET

The reproduction of part of employees' Time Table No. 9 depicts the intensity of service offered to Laurel Line patrons after June 14, 1909.

Edward S. Miller Collection

REAR ENDSHEET

The Laurel Line used an extensive variety of tickets between 1903 and 1952. A small portion of them are arrayed here.

Edward S. Miller Collection

Library of Congress Cataloging-in-Publication Data

Henwood, James N. J.
The Laurel Line.

(Special ; 103)
Bibliography: p.
Includes index.
1. Lackawanna and Wyoming Valley railway. I. Muncie, John G. II. Title. III. Series: Interurbans special ; 103.
TF725.L33H46 1985 385'.09748-36 86-909
ISBN 0-916374-72-6

TO

EDWARD S. MILLER

whose interest in the Laurel Line
is exceeded only by his warmth,
friendliness and generosity.

Acknowledgements

IN THE LONG PROCESS of researching and writing this book, the authors have incurred debts to many people who contributed material, photographs, suggestions, advice and criticism. Foremost among them is Edward S. Miller, who conducted the authors on a tour of the route of the Laurel Line, spent countless hours offering information and comments and freely contributed numerous photographs, maps, brochures, tickets and other memorabilia from his vast collection. In the course of many visits to Ed's home, the authors not only enhanced the quality of their book but also gained a friend as well.

The manuscript was improved by the thoughtful criticism and suggestions of Fred W. Schneider III and by the many leads he provided. Richard G. Prince also read every page and made comments and corrections. General advice and helpful aid were provided by Gerald E. Morris, Thomas T. Taber III and Francis J. Goldsmith, who also contributed a number of photographs from his collection. Several maps, artwork and the excellent car plans were drawn by Charles P. Yungkurth.

Photographs, so important to any work of this type, were kindly provided by Mervin E. Borgnis, Donald Duke, Thomas F. Flanagan, John Hoschek, Jr., and the Motor Bus Society, George Krambles, Francis P. Kutta, the late Stephen D. Maguire, Commander William D. Middleton, C.L. Siebert, Jr., Richard Short, Charles P. Yungkurth and the George Arents Research Library at Syracuse University. The colorful oil painting is the work of Frederick Beaver.

Thanks are also extended to Edgar Boland, a retired L&WV employee; David P. Morgan of *Trains* magazine; Frank J. Tobin, president of Ford, Bacon and Davis, Inc.; John H. White, Jr., of the Smithsonian Institution; the Westinghouse Electric Corporation and Albert W. Gendebien, archivist of Lafayette College.

Ms. Carolyn Davis and William H. Lee of the George Arents Research Library, Syracuse University, gave invaluable aid during the many visits the authors made to examine the extensive L&WV collection housed there. Other libraries and their staffs which proved helpful were Benjamin Kline, of the Pennsylvania Historical and Museum Commission, Railroad Museum of Pennsylvania; the Haverford Township Public Library; Kemp Library, East Stroudsburg University; the Lackawanna Historical Society, Scranton; the Scranton Public Library; and the Wyoming Historical and Geological Society, Wilkes-Barre.

Assisting in various ways were James J. Post and Deborah A. Wesley. Most of the typing was capably done by Mrs. Rose Reader; Ms. Sue Baldwin also helped perform this fundamental chore. To these, and to all others who helped in any way, the authors express their deep sense of gratitude. Any errors are, of course, our own.

Preface

THE DAWN of the twentieth century saw a new form of transportation evolve in the United States: the interurban electric railway. These enterprises were natural offshoots of the original, short urban trolley lines that quickly replaced the horsecar in the 1890s. Most of them lived in relative obscurity and followed a seemingly common pattern of promotion, construction, a few years of prosperity and possible expansion, and then an inevitable decline as they were beset with competition from newer forms of transport, changing lifestyles and financial pressures. A relative handful managed to survive until post-World War II years and thus have attracted greater attention. Among them was the Lackawanna and Wyoming Valley Railroad.

The Laurel Line, as the L&WV was most commonly known, fits the general pattern of intercity electric railways, but it was also unusual in several respects. It was built to higher than normal standards for electric railroads; it operated mostly with a third rail power system; it ran exclusively on private rights-of-way; it served a geographically narrow region whose economy was heavily dependent on one industry—coal; and, perhaps most importantly, many of its corporate records are extant.

It was this last element which enabled the authors to examine its life and to write its history. In doing so, we have tried to trace the corporate growth and development of the L&WV; to analyze its promotion, construction and operations; to examine its technology; to assess its management and daily routines; and to place it in its economic and social environment. In short, we believe the L&WV provides a case study of a small but diverse carrier which, while unique, also offers lessons about American railroading practices and styles in the twentieth century.

JAMES N. J. HENWOOD
JOHN G. MUNCIE
East Stroudsburg, PA
November 1984

Built too late, the Laurel Line was forced to bypass too many of the population centers already served by steam railroads. It cost too much to build and was perpetually saddled with a burdonsome debt. Ironically, its cars continued to ply between Scranton and Wilkes-Barre long after its competitors gave up. Eventually the Laurel Line also faced the inevitable. Determined not to let the vermillion passenger cars pass into oblivion without one last excursion, Edward Miller and William Ellis chartered this car for themselves and sixteen friends. On December 28, 1952 they photographed the last of many specials from across the ice-covered pond at Rocky Glen Park.

Edward S. Miller

Table of Contents

Illustrated is the Pennsylvania Coal Company's Number 6 incline plane in Dunmore, circa 1880. This distant predecessor of the Erie Railroad had two different routes between Pittston and Hawley. This view shows the eastbound or loaded track. The westbound or light track was out of sight to the right. Assemblages of cars were attached to a cable and hoisted up each incline. They were then allowed to coast down a less steeply graded alignment to the base of the next incline. Thus the technology of the era dictated the use of two different routes. Many years later the Laurel Line's Dunmore Branch was built on the right-of-way of the coal company's loaded track, which, in the interim, had been the Number 6 Branch of the Erie & Wyoming Valley Railroad, an immediate predecessor of the Erie. The Erie retained the light track's roadbed in this area for its own use. The Laurel Line's passenger shelter in Dunmore was located about 500 feet to the left of where the photographer stood. The other railroad on the bridge is the Delaware, Lackawanna & Western, which opened in 1856 and is used today by Steamtown excursions from Scranton to Moscow.

Dr. Edward Steers Collection

1

Building a Railroad: Laurel Line Style

IN THE FIRST HALF of the twentieth century, the Lackawanna and Wyoming Valley Railroad carried its passengers between Scranton and Wilkes-Barre, Pennsylvania, through valleys deeply scarred by man's efforts to extract anthracite coal from one of its few sources in the United States. Huge coal breakers dominated lesser structures clustered around mine shafts and slopes. These, in turn, were surrounded by stockpiles of timbers destined to support the roofs of underground tunnels, hopper cars for transporting the "black diamonds" to market, large piles of coal and dumps of culm. Spontaneous combustion had caused many of these culm dumps, consisting of mine waste, to explode and burn.

A heavy sulfuric gas covered most of the towns, combining with lead in paint and peeling it from houses and actually permeating the homes to discolor silverware and make life generally uncomfortable for the miners and their families. Mine water and human waste had turned the rivers into open sewers, killing off all life within them. Yet, the name selected by the Lackawanna and Wyoming Valley Railroad for publicity purposes—the Laurel Line—and the scenes on the covers of its advertising brochures recalled memories of the area as it had been hundreds of years before.

Early descriptions of the Wyoming Valley depicted a pristine area where any settler would gladly spend his lifetime. One observer indicated that "Never was the force of the earth so beautiful, so restful, so witching to the human eye. Mountains, promontories and gently rolling hills and restful valleys, all crowded with flowers, brilliant foliage, birds of song and silvery streams." The valley averages about three miles in width and the surrounding mountains are roughly 1,000 feet high in the east and 800 feet on the west. Wilkes-Barre Mountain brings the valley to an abrupt halt in the south, while the North Branch of the Susquehanna River enters the valley at Lackawanna Gap and then travels about 20 miles to the south prior to exiting at Nanticoke Gap. The Lackawanna River, which joins the Susquehanna near Pittston, begins in the beautiful hills of Susquehanna County. During the early period of settlement, it was ". . . a bright, clear stream of sparkling spring water . . . abounding in numerous varieties of fish, and in the early days swarmed especially with fine brook trout." The entire valley was heavily timbered with pine, hemlock and spruce.

This idyllic setting was inhabited by the Indians until early in the 17th century. Stephen Brule, traditionally cited as the first white man to descend the Susquehanna River, was employed by Champlain as one of the group exploring Lake Huron and eventually made his way down the river in 1616 or 1617 to Shamokin. Although some traders undoubtedly entered the area soon after Brule's trip, records of their activities are sketchy at best and it was not until 1737 that Conrad Weiser, an Indian interpreter, was sent by the government of Pennsylvania to meet with a council of the Six Nations at Onondaga that a documented, significant step was taken to open the valley to white settlement. In that year, he found two traders and three men in Wyoming, looking for land on which to settle. Moravians were quick to follow Weiser into the area to carry the gospel to the savages. The most famous of these was Nicholas Lewis, more commonly known as Count Von Zinzendorff, who came to Wyoming to preach to the Shawanese in 1742. He was not welcomed by the Indians since they believed he was interested in a silver mine, near which he placed his tent.

Both Pennsylvania and Connecticut claimed the area which Zinzendorff was attempting to Christianize. Pennsylvania's assertion was based upon an agreement with the Indians in 1726, while Connecticut believed it owned the Wyoming Valley as part of the "Western Reserve" included in its original 17th-century colonial grant. The New England colony believed it confirmed this position with the purchase of the area from the Six Nations of the Iroquois at the Albany Conference of 1754. Based upon the grant and the purchase, the Susquehanna Company of Connecticut felt justified in sending about 200 settlers to the Wyoming Valley in 1762. During the following year, the Delawares, encouraged by the Iroquois, attacked the settlement, killing 30 and driving the remainder back to Connecticut.

The two colonies continued claiming the valley, with Pennsylvania signing the treaty of Fort Stanwix in 1768 to procure a deed for the area from the Indians, and the Connecticut General Assembly officially designating it the County of Westmoreland in 1766. The issue was resolved in December 1782 when a commission set up by the national government met at Trenton and decided that Connecticut had no claim to the land. Militia from Northampton County took advantage of the decision in 1784 when they disarmed the Connecticut settlers, jailed 66 of them and placed Pennsylvania firmly in control of the valley. By 1811, the area was served by a postal route running from Wilkes-Barre, through Pittston and Providence to Abington, located to the northwest of the future site of Scranton.

Development of the Anthracite Coal Industry. Coal would prove to be the catalyst which would change the face of the valley. The presence of anthracite or hard coal had long been known in the valley, but it had not been developed, up through the Revolution, since it was almost impossible to burn. Shortly after 1790, Colonel Jacob Weiss convinced some businessmen in Philadelphia that the mineral had great potential and together they formed the Lehigh Coal Mine Company to develop surface mining around Summit Hill. Because of a continued lack of demand for anthracite through 1817, the founders of the company leased all of its lands, a total of about 10,000 acres, to Josiah White and Erskine Hazard, operators of an iron works in Philadelphia, and a third partner, George Hauto. These three men formed the Lehigh Navigation Company and began a program to improve navigation on the Lehigh River to provide a steady supply of coal to Philadelphia.

Very shortly, the Lehigh Coal Company merged with the Lehigh Navigation Company and eventually Philadelphia did begin to consume large amounts of anthracite. This market, along with other cities, encouraged the development of other coal fields. Increased productivity required expanded marketing facilities and the rivers proved inadequate for transportation, particularly in the Wyoming Valley, where the waterways did not flow to major northeastern cities.

Because of this need for a transportation system, the basically sylvan setting of the Wyoming Valley, now increasingly dominated by the tall wooden towers of the breakers and the expanding network of the mines, was altered further by the coming of the railroad. Unlike the two more southerly coal fields, the Lehigh and the Schuylkill, each of which was traversed by a major river offering a potential outlet to the big city markets of Philadelphia and New York, the Wyoming field presented a more difficult transportation challenge. The North Branch of the Susquehanna River, which flowed south from near Pittston, past Wilkes-Barre, eventually emptied into the Chesapeake Bay, thus providing access to Baltimore but not to Philadelphia.

To remedy this situation, the Commonwealth had undertaken a vast and extensive engineering project in the 1830s, which consisted of a system of canals and railroads designed to funnel the commerce of the region to the Quaker City. Part of the great enterprise was the construction of a canal paralleling the North Branch of the Susquehanna. Even before it was completed, however, the superiority and efficiency of the new railroad was amply demonstrated. The railroad not only could provide more direct and swifter travel than the slow and cumbersome canals, but it could also penetrate valleys surrounded by mountains where canals were not practical. Such was the case in the Lackawanna and Wyoming Valleys.

Coming of the Railroads. One of the first railroads in the region was chartered on April 7, 1832, by the Pennsylvania legislature as the *Liggett's Gap Railroad.* This carrier was empowered to build a line to Cobb's Gap on the east in the Pocono Range, which would connect with a road to be constructed from that point to the Delaware River near the Water Gap. The Liggett's Gap would run through what was later to be called Scranton, climb through its namesake gap to Clark's Summit, to a point on the New York State Line in Susquehanna County. It was also empowered to build a branch paralleling the Lackawanna River to its mouth near Pittston, where it could connect with the North Branch canal. Although nothing was done by the promoters of the Liggett's Gap Railroad, the discovery of iron ore, in conjunction with anthracite, led to the establishment by 1844 of what became known as the Scranton Iron Works, headed by the family whose name it bears. Eventually, the site also adopted the family name, and by 1860 some 9,223 people resided there. In 1880, the population had increased to 45,850.

Backed now by the enterprising Scrantonians, work on the Liggett's Gap finally began in 1850 and the name was soon changed to the *Lackawanna and Western Railroad Company.* The first trains between Scranton and Great Bend, Pennsylvania, on the *Erie Railroad,* ran in mid-October 1851. On the eastern end, nothing had been done by the *Delaware and Cobb's Gap Railroad,* but this line was merged with the L&W in March 1853 to become the *Delaware, Lackawanna and Western Railroad.* The line to the east, including a bridge over the Delaware, finally opened in 1856. Eventually, the DL&W constructed or controlled its own lines, extending from Hoboken, New Jersey, to Buffalo, New York, and numerous branches, including the Lackawanna and Bloomsburg, the northern section of which paralleled the western side of the Lackawanna River, and which connected with the *Pennsylvania Railroad* in Northumberland. The Lackawanna, with its headquarters in Scranton, would become a major connection for the Laurel Line; in time it would eventually absorb its smaller counterpart.

Even before the Liggett's Gap Railroad became a reality, the *Delaware and Hudson's Gravity Railroad* was in operation. A means had to be found to transport coal from the mines around Carbondale to the Delaware and Hudson Canal at Honesdale. John B. Jervis suggested a railroad which required an ascent of 950 feet to the summit of the Moosic Mountains and then a descent to Honesdale which lies 20 feet lower than Carbondale. The Jervis plan proposed five inclined planes to climb the western side of the mountains and three inclines on the downhill side, with stationary steam engines functioning on the ascending planes and a braking system on the descending planes. Steam locomotives would be required on the fairly level stretches between the planes on the eastern side of the mountains.

The canal basin received its first load of coal from the Gravity Railroad on October 8, 1829. By 1841 the line was carrying almost 200,000 tons per year, when the route was improved to handle the heavy volume of traffic. Because of the opening of new mines by the D&H, the Gravity Railroad eventually was extended south from Carbondale to Valley

Dickson Locomotive Company's plant served as a background for the normal activity in the DL&W's yard in Scranton about the turn of the century.
Lackawanna Historical Society

Junction in the town of Olyphant by 1858. Passenger service was available on the entire line by 1877, and the Gravity Railroad remained in service until 1899 when the canal was shut down. The limitations of the gravity principle were evident in 1860 when the D&H pushed four miles further south to Providence, a suburb of Scranton. Locomotive power was utilized on this section of the line.

Service to central Scranton was initiated in 1863 and a station and office building were erected at Vine Street and Franklin Avenue. In 1865, under the leadership of George Olyphant and Thomas Dickson, the D&H was moving nearly 5,000 passengers, as well as coal, over its combination gravity and standard locomotive system between Scranton and Honesdale.

Although the Gravity Railroad was successful, it had many limitations. Among them was the need to transship coal from the gravity cars to D&H canal boats, or in winter, to Erie Railway hoppers. Therefore, the D&H looked for a standard rail route for access to the coal fields. The Erie was also interested in a more direct connection with its main line, particularly for westbound shipments of anthracite. An agreement was made in 1868 between the Erie and the D&H, whereby the former would construct a line 37 miles long north from Carbondale to a point near Susquehanna on the Erie main line. The D&H was granted trackage rights over this route in return for financial assistance with construction costs. Chartered as the *Jefferson Railroad Company,* the line opened in October 1870. Shortly thereafter, in 1872, the D&H built a connection from the Jefferson Railroad to its main line to Albany, New York, known as the Ninevah Branch, giving the railroad a through route from Scranton to Albany. This route carried an increasingly greater share of the coal traffic of the D&H so that by 1900 the Gravity Railroad was abandoned.

Although the Erie, too, made heavy use of the Jefferson Railroad, it had an alternate route to the coal fields, which was developed much earlier. The Pennsylvania Coal Company, chartered in 1838, began construction 10 years later of a gravity railroad from Pittston to Hawley, 47 miles long, where it connected with the D&H Canal. Early in 1862, the Erie negotiated a contract with the Pennsylvania Coal Company, whereby it would transport the anthracite from the gravity railroad at Hawley to the various tidewater terminals. This necessitated the building of a branch from Lackawaxen on the main line to Hawley in 1863.

As with the D&H system, the gravity operation presented difficulties. Therefore, in 1864, the Pennsylvania Coal Company decided to construct a standard-gauge railroad on substantially the same route as the gravity line. Chartered as the *Erie and Wyoming Valley Railroad,* the line eventually included 75 miles of track with numerous branches. In 1884 the second gravity system was abandoned. Eventually, the Erie purchased a financial interest in the Pennsylvania Coal Company and the Erie and Wyoming Valley Railroad and by 1901 its control was complete. Thereafter, the Erie operated the line as its Wyoming Division.

The Steel Works

The Lackawanna Iron & Steel Co. blast furnaces (ABOVE) were viewed in 1892 from Front St. east of Pittston Ave. in Scranton. A small dam or boom held back the waters of Roaring Brook. The entire area became railroad property in 1901. The steel company locomotives (BELOW) were viewed in a small narrow-gauge yard under the Cedar Ave. bridge. These engines ran over rights-of-way later used by "modern" electric interurban cars.

BOTH PAGES: Lackawanna Historical Society

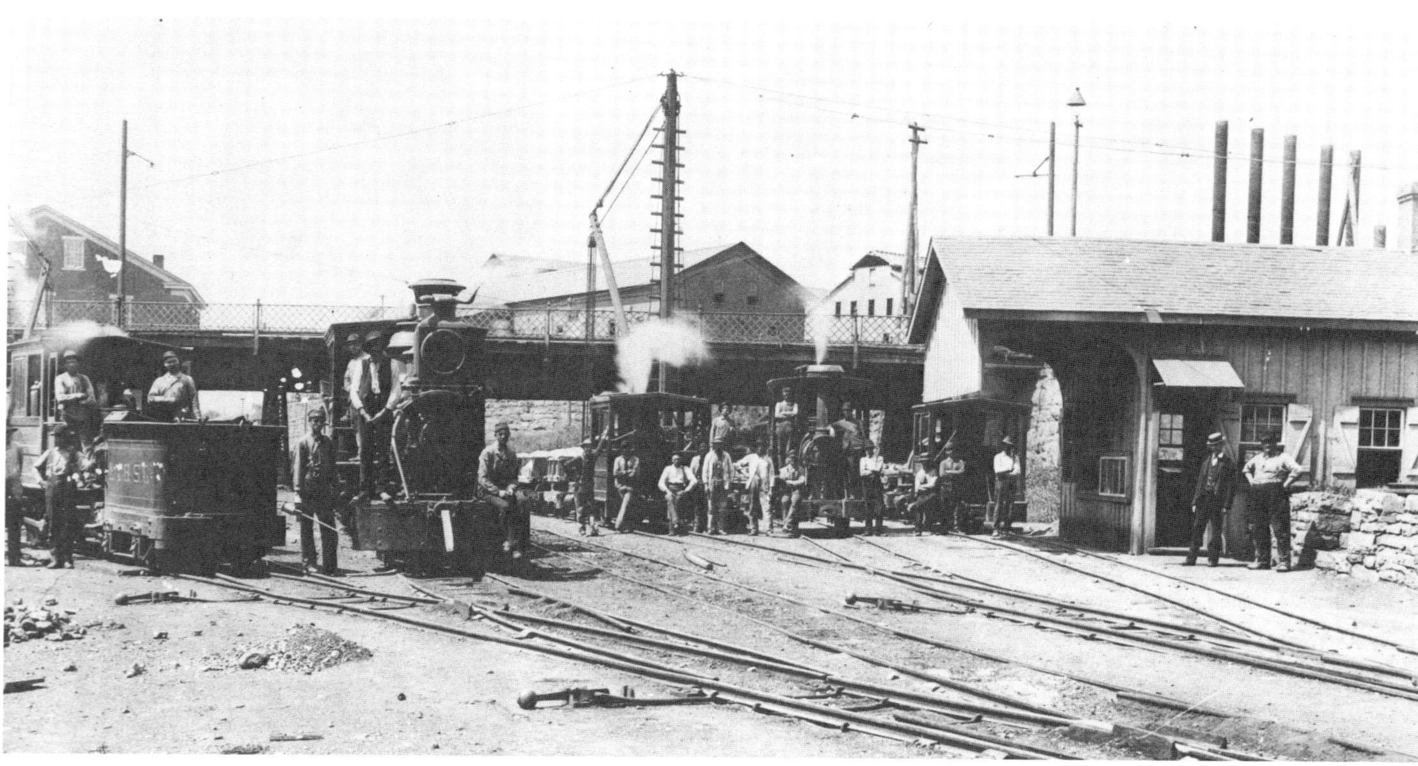

The entire valley of Roaring Brook seen in this picture looking east from the Spruce St. bridge was later deeded to the Laurel Line. The power house for the interurban was erected in 1902 on the site of the large center building; the carbarn rose to its left. The look-alike company homes to the left and right sides of the valley were owned by the steel company. The L&WV bought 98 acres of former steel company property.

A narrow-gauge steel company track climbed out of the valley to the ore roaster and slag dump. This site shows in the upper picture just to the right of the mills. The Laurel Line later used this right-of-way for its "over-the-hill" and Dunmore routes. Much later the Central Scranton Expressway connection to Interstate 81 was built on this very alignment. Today, the row of brick houses, on the west side of South Webster Avenue, remain in use except for the house on the left. It was demolished during construction of the Central Scranton Expressway.

Another carrier which entered the valley from the north, and one which would later figure in the plans and hopes of the Lackawanna and Wyoming Valley's managers, was the *New York, Ontario and Western Railroad.* Originally chartered in January 1866 as the *New York and Oswego Midland Railroad,* the O&W, as it was commonly known, ran from Oswego, New York, on Lake Ontario in a southeasterly direction through New York State to tidewater at Weehauken, New Jersey, which it reached over the tracks of the West Shore Railroad. Like other aspiring lines of the day, the O&W desired a direct entrance to the coal fields. This it was able to accomplish by acquiring various small companies and building connections.

The Scranton branch, extending 54 miles from Cadosia, New York, on the O&W main line through Forest City and Carbondale, Pennsylvania, opened on June 30, 1880. Attracted by the profits from the coal trade, the *New York, New Haven and Hartford* acquired control of the O&W in 1904. Around 1911, the company constructed a branch south into Old Forge where it connected with a branch of the Lehigh Valley.

As if the region were not already saturated with railroads, four additional carriers entered the valley from the south and east. One of these was the *Central Railroad of New Jersey,* which gained access by its lease of the *Lehigh and Susquehanna Railroad* in 1873. This company had been chartered by the Lehigh Coal and Navigation Company, one of the pioneer mining and transportation firms in the Middle Coal Field, in 1827. Nine years later it began carrying coal south from the Wilkes-Barre area. The lease of this carrier gave the Jersey Central, which crossed into Pennsylvania at Easton, a direct access to the coal region. By 1888 the Central built a branch north into Scranton, connecting with the Ontario and Western near the Lackawanna Avenue bridge in the center of the city.

A second line to penetrate the region from the south was the *Lehigh Valley Railroad.* Chartered in 1847 as the *Delaware, Lehigh, Schuylkill and Susquehanna Railroad,* the line was taken over by Asa Packer in 1851. He had the foresight to hire Robert H. Sayre as chief engineer and soon the road was under construction between Mauch Chunk and Easton. In

1853 the name was changed to the Lehigh Valley Railroad and under Packer's direction an aggressive policy of expansion was pursued. The company purchased a controlling interest in the old North Branch Canal from Wilkes-Barre to the state line in 1866, changed its name to the *Pennsylvania and New York Canal and Railroad Company* and proceeded to construct a railroad along its entire length. In 1868, Lehigh Valley trains entered Wilkes-Barre for the first time. Soon they were running to Pittston and beyond on the main line to New York State.

The giant Pennsylvania Railroad also entered the valley. Initially, the PRR became involved through its ownership of the Wyoming Valley Canal Company, which it acquired in 1869. This route extended from Wilkes-Barre to Northumberland, 64 miles. Various improvements were made and new boats purchased, but the business gradually declined and the operation was abandoned in 1901. In the meantime, the PRR had secured an all-rail route by its lease of the *Sunbury, Hazelton and Wilkes-Barre Railroad* in 1878. The new management completed the line to Wilkes-Barre by 1882, tapping various coal lands owned by Pennsylvania subsidiaries. Its northern terminus was the Lehigh Valley Station in that city, at a point directly opposite the eventual site of the Wilkes-Barre terminal of the Lackawanna and Wyoming Valley Railroad.

The last of the steam roads to enter the Wyoming Valley was the *Wilkes-Barre and Eastern,* an extension of the *New York, Susquehanna and Western,* which was opened to traffic in December 1893. The Susquehanna had an arrangement with the DL&W to carry coal from its terminus in Stroudsburg to a point near Hoboken, but it desired its own line directly to the coal fields. For this purpose, it chartered the Wilkes-Barre and Eastern, which ran from Stroudsburg over the severe Pocono grades to Hancock, later called Plains, directly north of Wilkes-Barre, where it crossed the Susquehanna River by means of a 1,200-foot bridge to Kingston. The Wilkes-Barre and Eastern, with its numerous steam railroad neighbors, blanketed the Wyoming and Lackawanna Valleys. Generally, they concentrated on coal and ignored the needs of local residents for swift and convenient transportation within the anthracite area.

Early Planning for the Lackawanna and Wyoming Valley. By 1900, then, the time was ripe for a more daring and innovative transportation enterprise. The valley region was growing and prosperous; the future looked bright. There were traditional railroads aplenty, but they were interested primarily in the lucrative coal traffic or the prestigious long-distance passenger market. Local service was slow and relatively infrequent and fares were high. Only the Delaware and Hudson Railroad offered a first-class service, scheduling 14 daily trains between Carbondale and Wilkes-Barre. Even here, however, coal traffic took precedence, and the line skirted the city of Pittston rather than operating through it.

It was possible to ride between Scranton and Wilkes-Barre by using the local trolley systems of each city. However, the cars were small and uncomfortable; the tracks ran through the streets of the numerous towns of the valley; schedules were slow and the ride was tedious. Few residents undertook the rigors of a through journey on the city streetcars.

The absence of good, swift, frequent transit service attracted the attention of promoters and entrepreneurs in New York and Philadelphia. It seemed to their creative minds that a high-speed electric railway, linking the valley towns from Carbondale on the north to Wilkes-Barre or even Nanticoke on the south, and offering both passenger and freight service, was bound to prosper.

The age of the streetcar arrived in Scranton in 1886, when Charles J. Van Depoele equipped a primitive electric line to the Green Ridge section of the city. By 1897, the trolley was the only fast way to get from one place to another in Scranton, except as shown on the facing page where the cars were mired down among the farm wagons and those of hucksters on Lackawanna Avenue's well-known wholesale block. *Lackawanna Historical Society* **On this page is illustrated Wilkes-Barre & Wyoming Valley Traction Company's parlor car *Anthracite*, on which invited guests rode in wicker chairs after 1893.**
Edward S. Miller Collection

The prime mover among the promoters was George A. Lee of Philadelphia, about whom little is known, except that he made a speciality of railway investments. Lee was one of the major forces behind the Transit Contract Company, which was formed to supervise construction of the line. Among the other promoters who became involved in the plan were U.S. Rep. William Connell of Scranton, a coal operator; Thomas F. Penman, the federal collector of Internal Revenue for the 12th District; Charles C. Mattes, the general manager of the Lackawanna Iron and Coal Company of Scranton; M.E. McDonald of Scranton, an attorney and "capitalist"; Matthew Stanley Quay of Beaver, a United States senator and powerful Republican "boss" of the Commonwealth; Thomas B. Simpson, petroleum producer of Oil City; State Sen. Benjamin K. Focht of Lewisburg; Henry C. Riley, a wholesale lumber dealer from Philadelphia; and George H.V. Martin, a banker from Camden, New Jersey. These and several other men set about to raise capital and to secure the legal rights to construct the proposed line.

Attracting investors in a region already covered with railroad tracks proved to be difficult. For this and other reasons, the promoters decided to charter several corporations rather than just one, although the line at its maximum was conceived as being only 40 miles long, from Wilkes-Barre to Carbondale. An amazingly complex series of transactions followed.

The first significant product of the promoters' efforts was the *Connell Park and Speedway Street Railway Company,* chartered on April 14, 1900, under the street railway code of the Commonwealth. This company planned to construct about eight miles of line. One section would extend from downtown Scranton in a southeasterly direction to the city limits; another would go in a northeasterly direction to the speedway located on the Nay Aug Falls and Elmhurst Boulevard, near the westerly boundary of Roaring Brook Township. Several loops and branches would complete the system. Essentially, the CP&S was to provide access to and egress from the city of Scranton. As eventually built, however, the Laurel Line would secure a much superior private right-of-way, bringing it to the heart of the city. With an authorized capitalization of $50,000, the CP&S would remain a paper corporation, useful only for raising money.

Then, on May 28, 1900, the *Scranton and Northeastern Railroad Company* was chartered. Despite its title, this line was to extend from the Scranton city limits in a southwesterly direction to the city of Pittston, about midway between Scranton and Wilkes-Barre, a distance of more than 11 miles. It was capitalized at $100,000.

Early in the following year, on February 14, 1901, the promoters applied to the state for a charter for the *Central Valley Railroad Company* to construct the line from Pittston to Wilkes-Barre. Among the organizers were Simpson, who was elected president, Riley, Focht, McDonald, Mattes and Penman. Initially, the company was capitalized at $100,000, but this was soon increased to $250,000. In addition, loans totaling $1,250,000 were raised by the company, bringing its total capitalization to $1,500,000. On April 12, 1901, an agreement was signed with Lee's Transit Contract Company, whereby that firm was to construct and equip the road for a fee of $1 million, one-fourth of which was to be paid in cash with the remainder covered by bonds. The Transit Contract Company, in turn, assigned its rights and fees to the engineering firm of Westinghouse, Church, Kerr and Company of New York on July 29, 1901.

To acquire an entrance into the city of Wilkes-Barre, on January 30, 1903, the Central Valley negotiated an agreement with the Lehigh Valley Railroad and its lessee, the Pennsylvania and New York Canal and Railroad Company, whereby the electric road would use the broad right-of-way of the canal from a point near the Susquehanna River to Pennsylvania Avenue near the center of the city. Since the cost of construction to date far exceeded the original estimates, it was necessary to apply for a charter for still another company, the *Wilkes-Barre Railroad Company,* on June 30, 1903, to purchase ground for a yard and freight and passenger depot fronting on Market Street, directly across the street from the Lehigh Valley Railroad station. This company was funded at $998,750 in stock and bonds and promptly leased to the Central Valley for 999 years, beginning July 1. While these activities were going on, the Central Valley itself was leased to the *Northern Lackawanna Railroad.*

On the same date that the Central Valley Railroad was chartered, February 14, 1901, the same promoters, with Thomas Simpson again serving as president, formed the Northern Lackawanna Railroad Company, to construct a line from Scranton to Carbondale. Capitalization in stock and bonds reached a total of $2,278,750. As it turned out, the only portion of the route to be constructed was a short branch from Scranton to Dunmore. On March 16, 1903, the name of this corporation was changed to the more descriptive title of the *Lackawanna & Wyoming Valley Railroad,* which leased the other firms and became the operating company for the electric line. All of the above companies, except the initial Connell Park and Speedway, were chartered under the steam railroad code of the Commonwealth, reflecting the ambitious aims of the promoters and the broader powers available to standard railroad corporations.

Westinghouse Gets Involved. George A. Lee and his Transit Contract Company served as the catalyst for this enterprise. Although five separate corporations had been chartered, the promoters conceived of this undertaking as a unified, high-speed, freight and passenger route, which would connect the valley towns and, hopefully, be a link in a newly conceived bridge route joining the west with New England.

According to his scheme, freight from the west would be routed to Wilkes-Barre via the Pennsylvania Railroad's Sunbury Line. The Lackawanna and Wyoming Valley would forward this traffic to Scranton, where it would be turned over to the New York Ontario and Western which, in turn, would interchange with the New Haven at Campbell Hall, New York. Using the Poughkeepsie Bridge, the New Haven would distribute this freight to various New England points. Needless to say, nothing came of these ideas, since the steam railroads had no reason or incentive to share their through traffic with a small electric line serving a local area. Nonetheless, the L&WV never really gave up hope of developing itself as a major freight carrier.

Another rumor, apparently circulated by the promoters, and appearing in the *Scranton Times* and the *Philadelphia*

George Westinghouse, one of the pioneers in electric railroads, was born in 1846 in Central Bridge, New York. By 1900 he had created an industrial empire built upon the invention of the air brake. Westinghouse Electric & Manufacturing Co. and Westinghouse, Church, Kerr & Co. were part of the empire. No longer part of the same corporate structure are WABCO, nee Westinghouse Air Brake Co., and Union Switch & Signal Co.
Westinghouse Electric Corp.

Public Ledger in August 1901 was that the new electric road would provide a direct connection with the Pennsylvania Railroad. This would give the giant carrier direct access to Scranton and a potentially large traffic in coal. It is doubtful whether the moguls in Broad Street Station seriously considered this possibility. But, as with the bridge line plan, this scheme would reappear periodically in future years.

All of this activity in the coal regions attracted the attention of the Westinghouse interests. Not only was George Westinghouse himself personally involved in electric railway experiments, but he had also established an engineering and consulting firm in New York City to provide expertise and planning for new enterprises and, hopefully, to funnel orders for equipment to the Westinghouse firms. This organization, Westinghouse, Church, Kerr and Company, evidently approached the promoters in mid-1901 and offered to bid on the engineering, construction and equipment work. According to George Lee, negotiations that were underway with several other engineering groups were terminated at the request of the Westinghouse firm, which assured the Scranton promoters that they ". . . would put the project through in the shape contemplated as quickly and with as little cost as could be done by anyone." Westinghouse decided to become financially involved as well, since the Security Investment Company, another of the Westinghouse firms, became a major investor in the new enterprise.

The reasons for Westinghouse's decision to become a major investor in the new electric road are obscure. The common belief is that he was interested in the line for experimental purposes. This seems unlikely, in that Westinghouse was busy developing an alternating current system, whereas the proposed Lackawanna and Wyoming Valley would be equipped with standard direct current apparatus. Furthermore, there is no evidence in the company's files of any experiments being conducted on the line. It is likely that the Westinghouse group had several aims, such as constructing a showcase line, featuring Westinghouse equipment, earning a profit from the anticipated revenues which the road would generate, and possibly as a site for experiments. In the prosperous economic climate of the day, new electric railroads were being constructed in many localities; the L&WV gave promise of being better than most.

Westinghouse, Church, Kerr and Company (WCK) in August 1901 proposed to George A. Lee, the president of the Transit Contract Company, that he make a complete audit of the accounts of his company and provide a statement of its assets and liabilities. Lee refused to do this, stating that WCK could examine his disbursements if they so desired. Under pressure from the Westinghouse firm, it was found that no books were kept by the Transit Contract Company. Lee explained, ". . . that it was considered unadvisable to have books that might be called for by the courts." Further examination by the Westinghouse firm revealed that vouchers for the purchase of the right-of-way were drawn in the names of individuals connected with the Transit Contract Company, rather than in the name of the company itself.

Despite these irregularities, Westinghouse decided to go ahead and an agreemnet was reached in November 1901 between the Security Investment Company, a Westinghouse subsidiary, and the Transit Contract Company. The latter company exchanged the bonds issued by the underlying companies for bonds issued by a new holding company, the *Lackawanna and Wyoming Valley Rapid Transit Company.*

George Westinghouse was one of the pioneers in the field of electric traction. Born in 1846, he had a natural mechanical aptitude. While perhaps best known for his invention of the air brake, Westinghouse was intrigued with the possibilities of using electricity as a mode of power in railroad transportation. He was fortunate enough to obtain exclusive rights to the patents of Nikolas Tesla, who was developing an alternating current motor. Westinghouse was to spend much time and energy in trying to perfect an alternating current system for electric traction. However, this did not stop his firm from producing and marketing direct current equipment, which was in common use on the early trolley lines then being constructed around the country.

The Lackawanna and Wyoming Valley, as envisioned by its promoters, would be built to mainline steam railroad standards. As such, it could have provided both a market for the products of the Westinghouse interests and the possibility for experimentation in the field of electric transportation. As it turned out, practical considerations dictated that ordinary direct current should be used and a third rail method of propulsion was decided upon, probably to save on feeder cables and substations, and in anticipation of heavy traffic.

Why Third Rail?

Wilkes-Barre & Hazleton car 204 was posed on the Bear Creek viaduct near West Hazleton in 1903 when this new third-rail line opened from Hazleton to Ashley. Service was extended into Wilkes-Barre in 1907. *Fred W. Schneider, III Collection*

Before explaining why the Laurel Line installed a third-rail power distribution system, it is necessary to define the term third-rail. All direct and alternating current systems require two or more wires or conductors between the energy source and the appliance or light connected to it. Thus the appliances in our homes have two-pronged plugs to connect them to the two wires to and from the power station. In typical electric railway practice, the two rails on which the cars ride serve as the ground or negative conductor between the locomotive or electric car and the power source. The ends of the rails were bonded with a copper wire to prevent sparking and loss of power. Generally, a single wire over the tracks was the positive conductor. Energy was collected from the wire through a trolley pole or some other sliding connector, and returned to the power house through the steel wheels and the rails. The Laurel Line used a trolley wire in congested areas. Elsewhere the positive conductor was a third or power rail, mounted about 20 inches outside one of the running rails and about six inches higher.

At the turn of the century, overhead trolley wire electrical systems were deemed inadequate for high speed or heavy duty railroad service. Even though a steel wire could conduct only 70 percent of the current carried by a copper wire of equivalent cross-section, the massive size of the steel third rail compared to a small copper trolley wire could transmit the heavy currents needed to move a long freight train or a subway train driven by up to 40 electric motors. In addition, railways powered from overhead trolley wires required heavy parallel feeder cables connected at frequent intervals to the contact wire to prevent inordinate voltage drops. The large cross-section of the steel power rail did not require feeder cables in most installations.

Perhaps, even more important, the small contact area between the collection wheel on the end of a trolley pole and the trolley wire was inadequate to handle the vast currents needed for heavy duty service. This was particularly acute at high speeds because wind tended to push the trolley pole away from the wire. One study in 1902 showed that a wheel could handle up to 1,000 amperes if the car were standing still but only 200 amperes at 60 miles per hour.

Power collection was not a problem on third rail systems, considering the vast contact area between a sliding shoe and the third rail.

Finally, it was argued that third rail systems were cheaper to construct and maintain, and provided more reliable operation. A decided weakness, which was downplayed by its promoters, was its vulnerability to severe weather, particularly in the northern regions where coatings of ice on the third rail could completely shutdown a railroad.

Between 1894 and 1907 in Chicago, New York, Brooklyn, Boston, and Philadelphia, third rail technology was applied to nine subway-elevated rapid transit companies. Two were opened in 1894 and 1895 using locomotives to haul trains of non-powered cars. The remaining seven properties were opened starting in 1898 using multiple-unit electric cars. General Electric or Sprague, one of GE's predecessors, engineered and equipped eight of those systems. Only the Brooklyn elevated lines were outfitted by Westinghouse. Although routes may have been rearranged, all of these systems remain in operation today.

From 1896 to 1906, steam locomotives were displaced by eight electrification projects, all using the third-rail system. Two were primitive and ephemeral projects in New England. One involved the West Shore Railroad from Syracuse to Oneida, N.Y. to permit the parent New York Central to ascertain the benefits of electrifying its entire mainline east of Buffalo. Fun and sun seekers from Philadelphia were able to ride electric cars over a Pennsylvania Railroad mainline from Camden to Atlantic City, N.J. Also included were three suburban commuter districts and one urban tunnel project. Seven of these projects were also outfitted by General Electric. The Long Island and New York Central suburban lines out of New York City persist.

Ill advised were twelve interurban railways built with third-rail technology between 1901 and 1907. All of these were so-built because their promoters reasoned that such power capacity would be needed to handle contemplated freight tonnage, heavy passenger loads, or extreme high speed service. All but two were General Electric projects.

Alas, it was only coincidental that not all were over engineered or over built. Most did not move interchange freight in sufficient volume to insure survival. Few operated either long passenger trains or extremely frequent accommodations. The Keeseville, Ausable Chasm & Lake Champlain opened in 1906 and used steam engines for its sparse operations after 1911. Seven more third-rail interurbans failed by the middle of the Depression. Typical of these was the Laurel Line's neighbor, the Wilkes-Barre & Hazelton, which seldom required the brawn of its sole electric locomotive. Often a single interchange freight car trailed an infrequent passenger car over the line between its namesake cities. Like so many interurbans, it expired of an overdose of carbon monoxide, a victim of the paving of the highway between Wilkes-Barre and Hazelton after 1930. By the time it closed in 1933, only four were left.

The Central California Traction Co. dismantled its electrical plant in 1946, yet it continues to dispatch freight trains behind diesels. The CCT was simply in the right place at the right time to profit from postwar industrial growth. The Laurel Line dragged on through 1952 simply because of its location in an urban connurbation. The Aurora, Elgin & Chicago, billed in 1902 as the fastest of all interurbans, carried commuters into Chicago in aging rolling stock until it failed in 1957 partly because of route changes necessitated by expressway construction. For two more years its short freight trains used up more cash than they generated.

But one interurban still operates. The Philadelphia & Western Railroad, conceived as a high speed route to Pittsburgh but built only into the rural area immediately west of Philadelphia, exists today as the Norristown High Speed Line of the Southeastern Pennsylvania Transportation Authority. Government subsidies, lightning-fast cars, and a prime suburban location have insured its survival. At issue, however, is a need to reduce liability exposure caused by accidental brushes with the third rail in a now highly populated area. A possible conversion to overhead trolley wire may be explored in conjunction with the purchase of new cars.

General Electric successfully promoted third-rail, direct current technology because Westinghouse virtually abrogated the field while immersing itself into refinement of single-phase, alternating current systems. The Westinghouse tests of alternating current systems go back to 1894, the same year that General Electric marketed its first third-rail system. This oblivious attitude was demonstrated in 1903 when George Westinghouse indicated he lacked knowledge about the Laurel Line and asked Westinghouse, Church, Kerr & Co. for information. Westinghouse was not even aware that his subsidiary was building a third-rail system at the same time they were striving so hard to perfect the alternating current technology.

The General Electric inspired third-rail system had been, in so far as interurban development was concerned, a flash in the pan. No effort had been made to develop slide shoes for trolley poles or other sliding contactors which, when used against a trolley wire, could have eliminated many of the problems associated with high speed, high capacity service under wire.

The Westinghouse alternating current system was first used commercially late in 1904 and very briefly placed the corporation back in the running with General Electric. After 1907 most interurbans designed for high speed or heavy duty service used the new Westinghouse system. By 1910 it had been discovered that alternating current was ideal for mainline railroad electrification projects but it had been found wanting for interurban railway service. Cars were extremely heavy and complicated, particularly when they also had to run on 600 volts direct current in cities. They were extremely costly to maintain. The most ephemeral was converted to high voltage direct current from overhead wire within 18 months. None lasted after 1925. But the third-rail systems had also found a niche. They remain standard today for almost all urban rapid transit projects.

Fred W. Schneider, III

The original Moosic station at Powder Mill Road was photographed in May 1903. In November the platform was moved to Spring Brook Road, about one mile north, as a result of a public petition. The photograph on page 148 looks in the opposite direction from the bridge. *Edward S. Miller Collection*

Although the promoters formed five separate railroad companies to raise capital, they recognized the need for an overall organization to run the lines as a single unit. For this purpose, they had chartered the Lackawanna and Wyoming Valley Rapid Transit Company in 1900 under the more lenient laws of New Jersey. Total capitalization was $6,500,000. The officers included President George C. Smith, who represented the Westinghouse interests; Vice Presidents Thomas B. Simpson; L.A. Watson of Scranton, president of the Title Guarantee and Trust Company and Elisha A. Hancock of Philadelphia; Secretary Dudley P. Phelps, second vice president of Westinghouse, Church, Kerr and Company; and Treasurer C.M. Vail. Serving on the executive committee were Smith, Simpson, Phelps, Henry J. Conant and George A. Lee. Additional Westinghouse influence appeared on the board of directors in the persons of Walter C. Kerr, vice president of Westinghouse, Church, Kerr and Co.; Charles A. Terry, secretary of the Westinghouse Electric and Manufacturing Co.; and George Westinghouse himself. It was evident that the Westinghouse interests would be well served by the corporate structure of the L&WV Rapid Transit Company.

The actual transfer of financial obligations was highly complex. It included an agreement whereby the Rapid Transit Company purchased the railroad properties of the Transit Contract Company for $11,500,000. To pay for this, the Rapid Transit Company delivered 64,750 shares of its common stock and authorized a bond issue of $10 million. The Rapid Transit Company, in turn, assigned the railroad properties to Westinghouse, Church, Kerr and Co. in exchange for "certain securities," and it issued $2,500,000 of 50-year, five percent, collateral trust bonds. The total value of all properties was estimated at $9 million.

The company recognized the need for a corporate identity which would appeal to the traveling public. Some newspapers referred to the line during its formative stages as the "Cannonball." This apparently was so popular in the valley that the management feared it would be impossible to change it. Despite this, the name *Laurel Line* was adopted to symbolize the common mountain laurel, which covered the hillsides through which the line would run. To provide a visual sign of the new enterprise, a contest was sponsored by the company, offering prizes for the best suggestions. Two winners each received $25 in gold. The final design comprised three T-rails, the ends of which formed a triangle, indicating the nature of the business and its method of propulsion. A wreath of mountain laurel surrounded the rails, symbolizing the line's triumph over its physical adversities. It was well that the new railroad had such a logo, for there were problems aplenty to overcome as construction began.

The L&WV crossed future U.S. Route 11 on this bridge. Dupont station was located at the north end of the bridge. Scranton Railway Company's trolley line to Pittston passes under the structure. Route 11 was later shifted to the Pittston Bypass to the east of this highway. The right (north) abutment remains to this day.
Edward S. Miller Collection

Constructing the Laurel Line. The man primarily responsible for construction of the new road and for its operations in its early years was Charles Fuller Conn. He was born in Concord, New Hampshire, on November 11, 1865. His first practical experience with railroads was a job as a brakeman for the Boston, Concord and Montreal Railroad in the summer of 1883. That fall he entered Dartmouth College and was graduated in 1887. His early experience on the railroad must have been appealing to young Charles, for during each summer vacation he worked again as a brakeman or conductor. His first full-time job was as a traveling agent for the Boston and Worcester Railroad in October 1887. He gained experience on steamships as a traveling agent or purser for the Providence and Stonington Steamship Company and later the Fall River Line. On November 7, 1889, he married Mabel S. Dwight. By 1896, he was the treasurer for the Boston Terminal Company, which was then building South Station.

In April 1901, Conn joined the engineering firm of Westinghouse, Church, Kerr and Company. He so impressed his superiors that he was selected to be the manager in charge of construction for the L&WV properties in October 1901. After the road opened, he was elected vice president and general manager beginning February 1, 1904.

Judging by his correspondence in the company files, Conn was an outgoing, flamboyant man, who was well aware of his position and the prerogatives it provided. He early developed the habit of seeking passes from the various major trunklines, offering in return to the officers of those companies a system pass on the miniscule L&WV. Needless to say, they were not impressed. Conn was often able, however, to secure passes for specific trips including not only himself but his wife and son, Dwight, especially when the latter enrolled in Dartmouth and required transportation to Hanover, New Hampshire.

Conn lived in the fine hotels of Scranton for part of each year and then rented an apartment in the prestigious Green Ridge section for the remainder of the time. He was active in civic organizations and the Scranton Country Club, and he maintained contact with his fellow Dartmouth graduates through participation in the activities of various alumni groups.

From the beginning, the L&WV was conceived as a high-speed, heavy-duty line, suitable for both passenger and freight traffic. Therefore, the most desirable right-of-way would embrace few road crossings, minimum grades and easy curves. Since the earlier steam railroads had already pre-empted the floors of the Lackawanna and Wyoming Valleys along the Lackawanna and Susquehanna Rivers, it was necessary to build the Laurel Line on the hillsides above the eastern bank of the Lackawanna River, before swinging down closer to the Susquehanna River south of Pittston.

The entire project was designed and supervised by Westinghouse, Church, Kerr under an agreement with the Transit Contract Company, which acted as general contractors and subdivided the work among various firms. The first step in construction was the acquisition of property for terminals, yards and right-of-way. As originally planned, the line was to extend from Wilkes-Barre, through Pittston and Scranton, to Carbondale, but as finally completed the main line ran from Wilkes-Barre to Scranton, with a short branch to Dunmore on the northern outskirts of the latter city.

In Scranton, the company secured a tract of 98 acres from the Lackawanna Iron and Steel Company in the heart of the city. The steel company was moving its operations to Buffalo and thus this prime land was available for development. The property extended about 2,000 feet from Irving Avenue to Spruce Street and had a frontage of 1,000 feet on Mattes Street, soon to be renamed Cedar Avenue. It varied in width from four to six hundred feet, with Roaring Brook, a tributary of the Lackawanna River, running through it. Not only was this property centrally located, but also it was immediately south of the Delaware, Lackawanna and Western Railroad station on Lackawanna Avenue. On this land would be located freight and passenger stations, the power plant, the carhouse, shops and yards. A connecting track would link it with the DL&W main line.

One problem with this location was that the only access to the south was over the formidable barrier of Moosic Mountain. This involved a four percent grade on both sides of the mountain. In order to circumvent this obstacle and to permit the carrying of freight, it was later necessary to construct a one-mile tunnel at considerable expense, which opened in 1905.

Leaving Scranton, the planned line extended up the mountain a distance of approximately one-half mile, where it reached a point known as Laurel Junction. Here, the mainline tracks curved to the south and the branch to Dunmore would turn to the north. A connecting track known as the Erie cutoff would link both routes, forming a large wye. The point where the connecting track joined the southbound main line would be known as Latches.

Dropping down the four percent grade a distance of about three-quarters of a mile, the line reached Connell Crossover, which would one day become Connell Junction, since at this point the line through the tunnel would join the main track. Three miles further, after passing Virginia and descending a 1.5 percent grade, the right-of-way skirted Rocky Glen, site of a new amusement park. Half a mile distant, down another 1.5 percent grade, the line served the Borough of Moosic and then headed toward Avoca and Dupont, climbing a 1.5 percent grade. South of Dupont the planned road dropped sharply, descending a 2.17 percent grade for about 1.5 miles, including a 12-degree curve to Pittston, the largest intermediate city touched by the L&WV, where the Erie railroad tracks had to be shifted and a creek relocated.

Leaving Pittston, the road encountered two very sharp curves of about 17 degrees and then made its way on moderate grades through the towns of Inkerman, Hilldale and Hancock, a distance of about seven miles, to River Street on the edge of Wilkes-Barre. Entrance to the city was on a 30-foot strip of the old canal bed of the Pennsylvania and New York Canal and Railroad Company. The terminal site of four and a half acres fronted on Market Street opposite the Lehigh Valley Railroad station and adjacent to the soon-to-be constructed terminal of the *Wilkes-Barre and Hazleton* railway, another third rail line. Sidings would serve various warehouses and industries and a track connection was made with the Lehigh Valley Railroad.

Land purchases began in 1901 with George Lee handling the bulk of the negotiations. According to the *Scranton Times,* by July 1901 more than half the route had been graded, ties were being procured and 90-pound steel rail for the mainline tracks and 75-pound steel for the third rail were on order from the Lackawanna Iron and Steel Company.

Various subcontractors were employed by the Transit Contract Company to do the work, but the prime one among them was John R. Lee of Paterson, New Jersey, evidently not related to George Lee. Agreements were negotiated with the Spring Brook Water Company, the Erie Railroad and its subsidiary, the Pennsylvania Coal Company, for rights-of-way and changes in bridges and tracks in Moosic, Pittston and Wilkes-Barre.

In Scranton, the company engaged a contractor from Philadelphia to clear the terminal property of the old steel company buildings, which covered approximately 300,000 square feet. The L&WV received $5,175 from the contractor for salvage. A peculiar problem with the Scranton terminal lands was the presence of coal under them owned by the Scranton Coal Company. Since the coal veins were already being worked, the coal company demanded payment for unmined coal to remain as support pillars for the powerhouse, car shops and stations. On April 10, 1902, the Lackawanna Iron and Steel Company and the Transit Contract Company agreed that the latter firm would pay approximately $25,000 for the 17,000 tons of coal under the three acres involved.

Construction generally proceeded smoothly, but there were occasional problems. The Scranton Department of Public Safety wrote to John Lee, one of the major subcontractors, complaining about numerous instances of flying rock and boulders damaging adjoining properties:

> A single piece of conglomerate rock . . . flew within the park . . . big enough to kill a dozen people had it been hurled at a time of day when people are apt to visit the park. I am besieged by neighbors because of the danger of flying rock. Unless you are more careful we will prohibit you from using dynamite.

In early 1902, the company began looking for crushed stone for track ballast. Initially, 100,000 yards were ordered, but much more was needed. Eventually, a subsidiary company, the Meadow Brook Crushed Stone Company, was formed to provide the material.

By the early spring of 1903, track construction was well underway between Scranton and Pittston, with workers ditching and trimming up. Standard steam railroad STOP, LOOK AND LISTEN crossing signs were placed at the few road intersections. Third rail warning signs were installed only where breaks in the rail occurred and people might be endangered by them. Overhead wire was erected at the car shops and in Wilkes-Barre from the gas works to the terminal for safety reasons. A telephone system was constructed paralleling the tracks for use in dispatching trains.

Local officials were assured that the work was of high quality and would greatly benefit the residents of their towns. For example, an official of the Transit Contract Company wrote with some exaggeration to the road commissioner of Plains Township, "When finished, it will be the most perfectly constructed railroad in the United States and you and the generations to follow you will have the satisfaction of knowing that you can journey on the safest, quickest and most comfortable road in Pennsylvania."

While construction was underway, the office force headed by Conn was busy handling the endless paperwork connected with the operation. During the formative stage of the line, it was decided the local office in Scranton would be responsible for determining that bills, prices and amounts were correct before forwarding them to New York. No actual books were kept in Scranton, only records of payments and the conditions of accounts. Confusion reigned under this policy with the New York office advising Conn to improve his bookkeeping methods. In January 1902, Henry Conant of Westinghouse, Church, Kerr, sent a detailed filing system memorandum for use by the Scranton office force. This was ignored and seven weeks later Conant sent another copy of the memorandum.

Since reports and paper continued to descend on New York in helter-skelter fashion, an exasperated Conant wrote to Conn in April pleading for his cooperation and pointing out that while it might seem to be a small matter, ". . . the magnitude of our correspondence is such that our business is almost impossible to conduct except by expending enough care in these details. . . ." Conant's ideas of "details" included a filing system of some 10 pages and hundreds of headings. It is no wonder that Conn chose to ignore it. This attitude was infectious and later managers also contributed to perpetuating a poorly organized filing system.

The lack of good filing procedures contributed to a practice that became habitual. Frequently, the line would question the bills of suppliers and shippers, while at the same time it was remiss in making payments. Numerous instances occurred during the life of the line. In April 1902, the company complained to the Lackawanna Railroad that its freight bill for shipping five cars of crushed stone was too high. "The price paid for the material itself was 25¢ a ton, consequently the freight rate of 1¢ a ton seemed to us excessive." There is no evidence the Lackawanna reduced its charges. At the same time, the American Bridge Company wrote a long series of letters complaining of nonpayment of its overdue bills.

At its board meeting in March 1903 the officers of the L&WV Rapid Transit Company made a progress report. The line between Scranton and Pittston was nearing completion and the powerhouse was almost finished. Extensive heavy work was underway south of Pittston, but little had been done beyond Dunmore on the proposed line to Carbondale. Since the entire bond issue of $3,400,000 had been spent and the line was only partially completed, a new consolidated bond issue of $10 million was prepared.

As completed between Scranton and Pittston, a distance of almost 12 miles, the road was double-tracked with crossovers at various intervals. A total of 199 tons of 75-pound rail and 2,679 tons of 90-pound rail was used. Stone ballast to a depth of six inches provided a base for the southern yellow pine ties, spaced 2,640 to the mile. Every fifth tie was six inches longer than the others to support the third rail insulating pedestals, which were placed between the double tracks to protect passengers boarding and leaving trains. At the few grade crossings, the third rail was broken and the ends connected by cables placed in wooden conduits. The ends of the third rails were flattened and turned down to release the shoe from contact and to reengage it smoothly. The width of the right-of-way varied from 60 to 150 feet, depending upon the terrain and was enclosed by a wire fence mounted on white cedar posts placed 15 feet apart.

No grade went much beyond two percent except on the line over the hill in Scranton. Only one curve exceeded 12 degrees and retaining walls were built in cuts where necessary. Fourteen steel bridges, varying from 26 feet to 93 feet, were supported on concrete abutments. In addition, at Avoca, the longest bridge on the line, measuring 682 feet, spanned the tracks of the Lehigh Valley Railroad, the Delaware and Hudson and the Scranton Traction Company. It consisted of nine deck spans varying in length from 50 to 102 feet with a maximum height of 60 feet. Total cost of the bridges on this portion of the line amounted to $274,584, with major spans at Spring Brook, Moosic and Avoca, costing in the vicinity of $25,000 each and the Avoca viaduct representing the largest expenditure of roughly $136,000.

The operating center of the Laurel Line was Scranton, where the powerhouse, shops and freight and passenger stations were located, and a yard extended in a southeasterly

A team of horses gaze from Tarbrake Farm at the nearly completed mainline of the L&WV at the future site of the Virginia crossover. Virginia station would later be built at the curve in the distance to the south. The fence on both sides of the right-of-way was a futile attempt to keep trespassers off the tracks.

Edward S. Miller Collection

direction about one-half mile, paralleling Roaring Brook. Two reservoirs were formed by dams to provide a water supply for the powerhouse, which was supplemented by city water.

The principal buildings were of Tudor Gothic type made of brick with concrete sills, cornices and copings. The two-story passenger terminal contained the usual facilities for the traveling public and the company's offices. A rear platform adjoined the loop which turned the cars. A separate freight station was nearby, with a long siding and a roadway extending to Cedar Avenue.

Fifty feet above the yard and near the DL&W tracks were six company houses of varying sizes, which could accommodate 17 families. They were located on a private street and each had front and back yards. They were rented to employees who might be needed quickly, particularly during snowstorms or emergencies.

The carbarn, near the south end of the yard, was capable of housing 27 cars in three bays and contained the latest equipment for heavy repairs and servicing. The adjacent powerhouse contained five modern boilers fed by stokers, two Westinghouse 2,000-horsepower engines, two AC/DC generators and four transformers. Copper transmission lines carried the AC voltage to the substation at Hancock, where it was reduced and converted to a direct current of 650 volts and fed to the third rail. The building also contained passenger and freight facilities. Pittston had separate passenger and freight structures, as befitted its status as the major intermediate city on the line.

The southern terminal of Wilkes-Barre, the seat and commercial center of Luzerne County, received a substantial

This crossover was installed to permit cars leaving the Scranton terminal to reach the car house. Clearly shown is the complex third rail construction needed to prevent a power interruption when cars passed through the switches. Spring frogs were used to provide a smooth ride.

Edward S. Miller Collection

brick station on Market Street, with a loop track in the rear, and a small separate building used as a trainmen's room.

The other shops on the line were served by small shelters or open platforms. George Smith, president of the company, preferred platforms with a length of 100 feet but this was rejected by the Westinghouse officials in New York. Maple Street, for example, in South Scranton, had a wooden platform 60 feet long and 9 feet wide.

A dream long held by the promoters but one which was destined never to materialize was the plan to extend the road north from Scranton to Carbondale, a distance of 17 miles. The only part of this extension to be constructed was the short branch to Dunmore, one and a half miles distant. Unlike the main line, which had 90-pound rails, the trackage to Dunmore consisted of 75-pound rails. A total of 718 tons of steel was used for this extension.

This route left Scranton terminal on the same line up Moosic Hill used by trains to Wilkes-Barre. As described in an early history of the road:

> Leaving Scranton terminal . . . the route skirts the rocky gorge at the bottom of which, 150 feet below, Roaring Brook winds its way toward the Lackawanna River. Looking back towards Scranton as the road rises, several magnificent views may be attained of the city with its background of hills and the river winding through the valley below.

At the top of the hill, at Laurel Junction, the Dunmore branch swung to the left in a northerly direction. Completing its broad turn, the line passed Nay Aug Junction where a connecting track, known as the Erie cutoff as it was a part of the number 6 branch of the Erie and Wyoming Valley Railroad, ran to the south and joined the main line at Latches. The Dunmore line itself continued to the borough of that name.

The line from Laurel Junction to Dunmore, 1.01 miles, was not owned by the L&WV but was leased from the Erie Railroad's subsidiary, the Erie and Wyoming Valley Railroad. On October 27, 1903, the L&WV signed a lease with the Erie and Wyoming Valley for a term of 900 years. Under its provisions the lessor was to maintain the property and pay a sum of $40,000 in annual installments of $2,000, with five percent interest on the unpaid balance.

Beyond Dunmore, a preliminary survey in 1900 projected the line to Carbondale, passing through Throop, Smoketown, Dickson City, Olyphant, Mott Haven, Peckville, Archbald and Jermyn en route. Lack of money prevented the L&WV from carrying out these plans. In 1905, George Briggs, a local surveyor, in a report to Charles Conn, noted many obstacles which had appeared along the route. Among them was a road used by the Scranton Gas and Water Company, refuse from a coal washery near Throop, newly constructed houses within the boroughs of Throop and Dickson City, coal mines near Olyphant and Mott Haven and a washery and railroad spur at Archbald.

The failure of the L&WV to initiate construction on the Carbondale line aroused the ire of George A. Lee. In a letter to George Smith in January 1905, Lee complained about the long delay and urged that construction begin immediately in the spring. Lee felt the line would generate extremely profitable business and would put pressure on the steam railroads, which had still not consented to interchange agreements with

The Scranton terminal and loop track was partly completed during the winter of 1902-1903. Sufficient third-rail had been installed to permit testing of cars as they were received but the southbound mainline track remains incomplete. A temporary interchange track to the DL&W permitted delivery of supplies; it passes through the later site of the station excursion platforms. The DL&W passenger station was a half-mile farther west along Lackawanna Avenue at this time; the present building would be erected to the right six years later. *Edward S. Miller Collection*

the L&WV. He was particularly concerned about the supposed purpose of the L&WV:

> It is perhaps not inopportune at this time to also call to your attention a fact that has embarrassed us in more than one direction . . . , namely, that in attempting to arrange other plans on our own account we have been met with the statement which seems to have had more or less wide circulation that this road was built by your interests for the sole purpose of demonstrating not only what your interests could do in the line of constructing a high speed electric road, but also what a high speed electric road should be when properly constructed. We believe that you will realize . . . the difficulty and embarrassment such a statement as that might cause us. . . .

Evidently Lee was miffed in that he had promoted the line as a highly desirable business venture, which would generate large profits. The failure of the company to do so and the hostility of the steam railroads, along with the popular belief that Westinghouse was interested in the line only for experimentation and his own purposes, frustrated Lee. Nonetheless, no steps were taken to begin construction to Carbondale.

Despite the lack of action, the idea of an extension did not die easily. One plan that was briefly considered was the construction of a tunnel under the DL&W tracks in Scranton, which would avoid the route over the hill. In 1904, a rumor circulated to the effect that the L&WV was planning a line to Sunbury and Harrisburg. This was promptly denied by Conn,

The Power House

The heart of L&WV operations was the power house along Roaring Brook in Scranton. Heading two ash cars in the exterior view is number 10, one of two steam engines that ran on the railroad and the only one owned by the L&WV. Retaining walls and mill foundations remain from Lackawanna Iron & Steel Company's use of this site. The interior views, clockwise from upper left on the facing page, include one of the two original generators, the steam driven direct current exciters, the boiler room, and the direct current switchboard. The boiler room originally housed five Babcock & Wilcox 400-horsepower watertube boilers, one of which was a spare. The pair of 2,000-horsepower generators, one being redundant, were driven by Westinghouse vertical cross-compound steam engines with Corliss valve gear. Both of these units were dual generators which produced both 390 volts three phase alternating current and 650 volts direct current. The three phase current was stepped up to 22,000 volts and transmitted to rotary converters at Hancock and later at Avoca. The direct current was fed directly to the third rail at Scranton. The field coils on whichever generator was running were charged by one of the steam driven exciters in order to insure a constant voltage. All of these exposures were made when the plant opened in 1903.

ALL: Edward S. Miller Collection

and it was highly unlikely in any event, as the company could not even build the much shorter route to Carbondale.

In 1907 another idea surfaced. This was based on a preliminary study which considered the electrification of the Delaware and Hudson Railroad from Scranton to Carbondale. Under this plan, a track connection would be built between the L&WV and the D&H in Scranton and two of the main tracks of the latter company would be electrified between that point and Carbondale. The L&WV would be granted trackage rights to run its trains to Carbondale on this line. Although Smith wrote to C.S. Sims, vice president and general manager of the D&H, encouraging him to consider this matter, the D&H was not interested. In 1908 consideration was briefly given to an extension from Carbondale to Cortland and Syracuse, N.Y. This plan was impractical for the limited population along the route was already served by other roads. Fortunately for the L&WV, this scheme never materialized.

In 1909 still another idea appeared. A Mr. R.W. Day of Scranton proposed the construction of an electric railroad from Laurel Junction in South Scranton to Moscow, Lake Ariel and Honesdale, a distance of about 38 miles. He suggested the new company use the L&WV trackage down the hill and the Scranton terminal and the L&WV supply the power for the projected line. Although the population in that area was sparse, Day evidently hoped to develop a summer trade to Lake Ariel, since he expected to need two and three-car trains on frequent headways.

Conn was intrigued with the idea and recommended it favorably to Smith. He was unconcerned about the Erie Railroad's line to the lake, since that company was not interested in developing the passenger business. Conn also felt the new electric road could divert passengers traveling from Honesdale to New York from the Erie branch to Lackawaxen, where it connected with the main line, to Scranton and the DL&W connection. Smith indicated to Day that he would favorably recommend to the board of directors a long-term contract with his company and he spelled out the details of the arrangement. Wiser heads prevailed and the proposed electric line never advanced beyond the speculative stage.

By 1910 the plan of extending the line to Carbondale was dropped. A buyer in that city offered $3,000 for the station site, which the L&WV had purchased in 1901 for $4,800. Smith felt that if the price was fair, the land should be sold. Although occasional proposals would still appear, for all practical purposes, any extension of the Laurel Line was dead.

The L&WV planned the "over-the-hill" mainline through South Scranton as a temporary line, to be used until a tunnel could be built. The substitute nature of the line is most apparent in the sharp curve where the mainline swung south from Laurel Junction and around the former steel company slag dumps. This curve was the sharpest anywhere on the railroad between the terminal loops at Scranton and Wilkes-Barre. It was in marked contrast to the high speed curve on the Dunmore branch in the foreground, which was conceived as a permanent route to Carbondale. The planned northern line, which here passes through the foundations of the old ore roaster, was opened to Petersburg, 2.24 miles from Scranton and less than a mile north of this junction, on June 20, 1904. What turned out to be the final quarter-mile to Dunmore was placed in service June 27, 1905. To the far left is the former Erie & Wyoming Valley Number 6 Branch, leased by the L&WV to provide a connecting link between the Dunmore branch and the mainline. It was used for scheduled passenger service from 1914 to 1945. To the right of Roaring Brook, a Delaware, Lackawanna & Western freight train climbs toward the Nay Aug tunnels. Other views of this same locale appear on pages 126, 146 and 186.

Lackawanna Historical Society

One of two Whiting 15-ton overhead cranes was set up to lift coach 7 in this 1903 view of the Scranton car shop. Note the lifting beam placed under the near end of the car. The horizontal light colored band on the car trucks just above the floor is part of the Gibbs cradle-suspension system of mounting traction motors. The carpenter shop had turned out two new wooden pilots for the two cars then in the building. *Edward S. Miller Collection*

Initial Rolling Stock. Of early concern to the managers was the ordering of rolling stock. The initial equipment was furnished by the American Car and Foundry Company's Jackson and Sharp Works at Wilmington, Delaware. One additional car, No. 301, used for construction and work purposes, was built at the AC&F plant in Berwick, Pennsylvania. The order included 10 single-end passenger cars, designated type L and numbered 1 through 10, costing $16,000 each; four combination passenger–baggage cars, originally numbered 11–14, but soon changed to 101, 102, 103 and 104, costing $16,000 each; four express cars, numbered 201 through 204, costing $9,000 each, and one work car, No. 301, costing $8,000.

The dimensions of the coaches and combines were almost identical. They were 42 feet long over the end panels and 52 feet long over the anti-climbers, with a width over sills of nine feet and a height from the bottom of sills to top of the roof of 10 feet. The platforms on each end were five feet. The distance from the top of the rail to the first step was 10 inches; five eight-inch steps led to the car floor. The cars were painted chrome orange and lettered RAPID RAILWAY in gold on the sides and ends. The passenger carbodies weighed 32,000 pounds and the combination cars weighed 35,000 pounds. Trucks were supplied by the Baldwin Locomotive Works. They were MCB swing motion equalized bolster type. Mounted on the rear trucks were two single reduction 50F Westinghouse motors of 150 horsepower each, probably operated by Westinghouse drum controllers. Steel axles, cast steel wheels and steel tires with MCB flanges completed the trucks.

The interior of each passenger car was divided by a wooden and glass partition into smoking and non-smoking sections; the former seated 25 and the latter 36. Double sliding doors were located in the end and middle partitions. When closed, the vestibule doors were secured by the lowering of a trap door, forming the extension to the platform covering the steps. Double-thick American glass windows were protected by two polished brass rods running the full length of the outside of the car. Dark red pantosote curtains operated on springs. The red plush seats were stationary and the cars were finished in natural mahogany inlaid with holly. Green opal glass was fitted into the fan lights in the clerestory. Eight electric lights of 16 candlepower on each side and one light in each vestibule illuminated the cars. Each coach had 20 electric heaters built by the Consolidated Car Heating Company.

The combination cars had a passenger compartment seating 44. The baggage compartment was 12 feet nine inches long with sliding doors on each side. Drop seats at the sides held 13 people.

The express cars had a body length of 46 feet, six inches, a width of nine feet, eight inches, and were built with straight grain, yellow pine with oak sills, strengthened with one-half-inch angle iron and oak cross timbers. Five-eighths-inch white wood, tongue and grooved 1¾ inches wide, sheathed the interior. Like the passenger cars, they were also single-ended and carried the same electrical equipment and heaters as the other cars. A sliding door was located in the middle of each side and there was a rear door as well.

The work car was a standard flatcar with movable sides 40 inches high. It had wooden pilots on the front and rear trucks, each of which had two No. 76 Westinghouse motors of 75 horsepower, with single reduction gearing to the axles. Unlike the other cars, it had chilled cast-iron wheels.

All the cars were equipped with trolley poles for operation in the yards and in Wilkes-Barre, as well as third rail shoes. A wooden locomotive type "cow catcher" was attached to the front under the link and pin Van Dorn coupler attached to standard radial drawbars. Westinghouse automatic type air brakes were supplied by type D motor-driven reciprocating air compressors located under each car. There was a vertical hand wheel brake in the motorman's compartment.

While the cars were being built, Westinghouse, Church, Kerr asked Conn to have someone research the local history of the valley region for the purpose of finding names for the cars. Indian names, such as Lackawanna, Wyoming, Luzerne, Nescopek, Nay Aug and Susquehanna, were suggested. Conn sent the list to New York, but it was rejected because some of the names submitted were not local in nature. He was told to do further research. Among the names finally chosen were

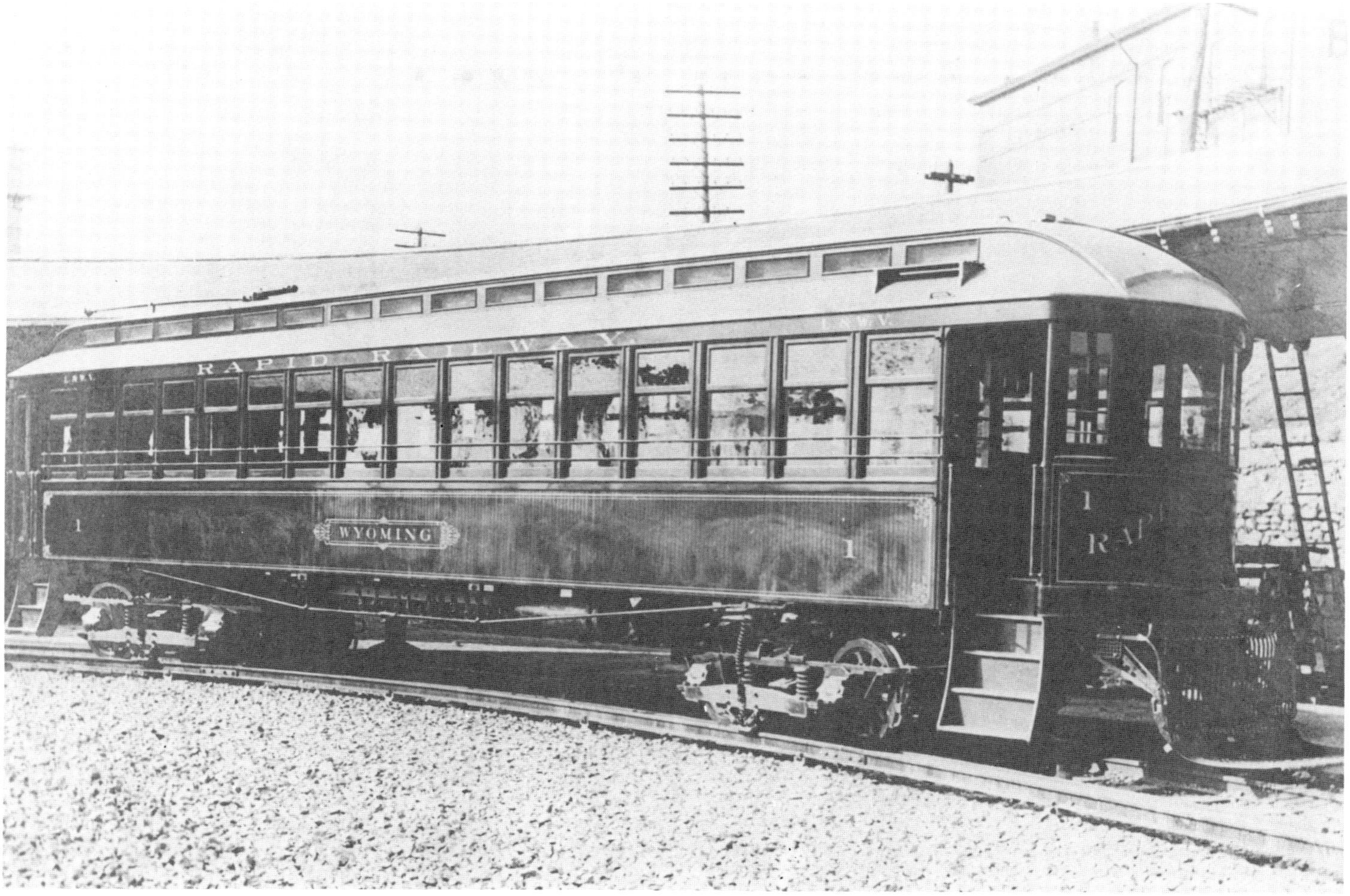

This photo of attractive wooden coach number 1, the "Wyoming," is rare for it is one of two known photos of a Jackson & Sharp car on the L&WV property which was equipped with the original fender on the front. These interurban-style life saving devices were replaced with pilots styled after steam locomotive "cow catchers" before the cars entered service. Reasons are unclear but it is probable that the original design swung over the station platforms in Wilkes-Barre and Scranton, presenting a hazard to waiting passengers. As built the Type L cars had doors front and rear on the right side but only a rear door on the left side. In order to take this traditional platform side view of the car, the photographer created another unorthodox situation — one of the few known photos of a car facing the wrong way on the Scranton terminal loop. The first two cars were delivered November 6, 1902; this view during testing was made February 10, 1903.

Lackawanna Historical Society

The photograph of the Baldwin six-foot wheelbase motor truck under one of the Type L cars depicts a host of engineering practices no longer in vogue. The steel tires were detachable from the spoked hubs in order to minimize replacement costs. Note also the sand tank mounted on the truck, the canvas conduit for electric cables under the floor, the gravity operated third-rail collector shoes, and the truss rods that kept the wooden car frame straight. *Edward S. Miller Collection*

The interiors of the original cars were spartan by modern standards but not unlike railroad coaches produced in the same Wilmington, Delaware factory. This was style in 1903 — red plush upholstery, or leather in the smoking section, mahogany walls with holly inlays, green opal glass in the clerestory, and two amenities far better than the railroads had to offer: dim bare light bulbs instead of dimmer oil or acetylene light fixtures, and electric heaters instead of coal stoves. Shown are both a coach (ABOVE) and a baggage-passenger combine (BELOW).

Edward S. Miller Collection

The handsome 4-4-0 locomotive "George A. Lee" sits on a rock crusher siding in the summer of 1902. The operation later became the Meadow Brook Crushed Stone plant near Connell Junction in South Scranton. Details of the locomotive appear in the rosters at the rear of the book under number 4.

Edward S. Miller Collection

The passenger car "Wampum" had already been renumbered from 13 to 103 in anticipation of the delivery of a type M car numbered 13 when this view of the Scranton terminal was taken in the summer of 1903. The longer of the two wooden cabinets in front of the car housed headlights and lanterns that were hung on the cars for night runs. And behind those striped awnings are a host of items not savored in a modern office building. Remember when paper weights never seemed to be in the right place when a gust tore through the office? Pens that dripped? Ink wells that invariably turned over on top of a completed ledger? The host of typists banging away on manual machines to produce multiple copies with carbon paper? Ah . . . The Good Old Days. *Lackawanna Historical Society*

The main waiting room in the new Scranton terminal awaits the influx of passengers in 1903. In an era of cheap labor, it was possible for the Laurel Line management to specify hand-laid, one-inch square mosaic floor tiles and a highly varnished, oak tongue-and-groove board ceiling. The room was illuminated by a row of small incandescent bulbs around the perimeter and a single row on a north-south axis in the center. The ticket office extended into a bay window on the loop side of the building. Four polished brass spitoons served a practical function in a less sanitary era. Note the permanent radiator in front of the center pair of doors on the street side — the doors are not false and an examination with a magnifying glass discloses hinges and a lock!

George Arents Research Library, Syracuse University

Wyoming, Nay Aug, Onas, Oneida, Capoose, Luzerne, Wampum, Westmoreland, and *Powhaton.*

For freight service, 10 standard boxcars, numbered in the 500 series, and four flatcars, numbered in the 600 series, were ordered. The flatcars were former DL&W cars, 32 feet long by 8 feet wide of 40,000 pounds capacity. They were acquired from AC&F at Berwick for $250 each. Initially, freight service was handled by steam locomotives.

The contractor used at least two steam locomotives. Number 4, a 4-4-0 type, was originally a Lackawanna and Bloomsburg engine, which was purchased by the Lee Construction Company in March of 1901 and was used in the building of the line. It was given the name *George A. Lee.* The other locomotive utilized in construction was No. 5. Neither of these engines remained on the property after construction was completed.

The L&WV, however, did maintain a steam locomotive, at least until it acquired an electric locomotive in 1906. This engine was formerly *Morris and Essex* No. 102, built by the Dickson Locomotive Works in 1873. It was a 2-6-0 type, weighed 98,000 pounds and had 54-inch drivers. It became DL&W No. 244 and was sold to Westinghouse, Church, Kerr in June 1902. As such, it assisted in the construction of the line. The L&WV purchased it for $3,600 and renumbered it 10. For a few years, it was used in freight service, supplementing the express cars

Winters often can be severe in northeastern Pennsylvania. Despite this, the L&WV never felt it was necessary to buy snowplows or sweepers. In December 1903, Conn estimated the cost of a rotary plow at $16,000. A cheaper means of handling snow was to apply a flanger to a motor car, which could be done for about $225, although no action was taken at this time. Sleet scrapers and brushes were used to clear the third rail of ice and snow.

Opening the Laurel Line. By May 1903, all was in readiness for the formal opening of the section of the road between Scranton and Pittston. As had been agreed, the Transit Contract Company would operate the line initially until it was formally completed and turned over to the L&WV Railroad. Members of the local press were ecstatic about the new railroad in the valley. The *Scranton Tribune* hailed the carrier ". . . the finest electric road in the United States. . . ." On Monday, May 18, about 50 newspapermen from the region were treated to an inaugural ride. They were accompanied by Conn, the vice president and general manager; B.F. Wyly, Jr., traffic manager; Charles P. Wilson, superintendent; and Major T.F. Penman, one of the directors. The trip was run in leisurely fashion, the explanation being given that the newness of the roadbed and equipment precluded a high-speed run.

However, a few days before, a car carrying only company officials covered the distance of nearly 12 miles in 24 minutes, including five stops. On the return trip, the reporters were much impressed by the power in the third rail, when a workman accidentally dropped a 1¼-inch chain across it. The chain melted in two, burning ". . . for six inches on either side of the third rail as if it were so much paper." Regular service began at 6:00 A.M. on May 20, with cars operating on a 20-minute headway and taking 50 minutes for a round trip. A temporary wye behind the Pittston station was used to turn the cars. The crowds were so great that ticket sales had to be suspended in midafternoon.

Not everybody obeyed the photographer's injunction to hold still. One lad in knickers in front of the Pittston station's bay window moved during the long exposure causing a ghost image back in the summer of 1903. Because the line was not yet open south of Pittston, car 103 would cross Market Street, switch to the north-bound track, and then back into the temporary wye in the background. This procedure was used to reverse the single-end car for the return trip to Scranton.
Lackawanna Historical Society

The *Tribune* felt the new line would more than compensate for the closing of the Lackawanna Iron and Steel Company's plant. It confidently asserted that the L&WV would employ ". . . day in and day out quite as many men that were employed sporadically by the steel company." In a glowing editorial, the *Tribune* predicted that the electric road would be a major element in the growth and development of the valley towns. Travel time between Scranton and Wilkes-Barre would be reduced by one-half and would be cheaper than the older means of transportation.

Somewhat provincially, the *Tribune* claimed that Scranton would gain "the lion's share" of benefits as the leading city in the area. "The completion of this great enterprise . . . is a subject for the liveliest congratulations. May its projectors meet with the reward in patronage which their broad-minded policy merits." While the promoters may well have gained substantial rewards, time would show that the stock and bondholders had not.

Work continued on the southerly portion of the line between Pittston and Wilkes-Barre. On September 15 service was extended to Hancock, where passengers could continue their journey on the slow trolley cars of the Wilkes-Barre and Wyoming Valley Traction Company. A temporary wye at the Hancock station permitted the cars to turn for the trip north. So great was the desire to extend operations that wooden bridges were installed at several points until the permanent steel bridges arrived from the builders.

Construction was particularly difficult for the remaining miles to Wilkes-Barre. A cut 1,000 feet long, 40 feet deep and 30 feet wide had to be opened below Hancock. The second largest bridge on the line, 554 feet long, spanned a roadway and two railroads, the Lehigh Valley and the Jersey Central, at the Prospect Colliery in North Wilkes-Barre. Public highways, streetcar lines and the Wilkes-Barre and Eastern bridge over the Susquehanna had to be modified to permit the passage of the Laurel Line. Thousands of feet of rip-rap were dumped to protect the roadbed from the Susquehanna River.

In Wilkes-Barre, the Lehigh Valley Railroad was busy moving its tracks in the old canal bed to make room for the new electric line. Overhead trolley wire was erected because of the several grade crossings and work was well underway on the new Wilkes-Barre station. The first car ran from Hancock to Wilkes-Barre on December 14, 1903. Regular 20-minute service began two days later with a scheduled running time of 43 minutes. The official opening of the complete line was held on Christmas day when more than 13,000 passengers were carried.

As the tracks progressed, other services were added. Beginning August 1, 1903, express service at freight rates was offered between Scranton and Pittston. The line was very interested in handling carload freight but the severe grades over the Moosic hill, which would be eliminated in 1905 by the tunnel, and the hostility of the steam railroads to an interchange agreement, prevented any action in that area. The first chartered car was operated on June 8 for Miss E.L. Wilcox and party. Record crowds were carried on October 29, a miner's holiday commemorating the accomplishments of

United Mine Workers' President John Mitchell. Of the 13 available cars, 12 were placed in service and one held for emergencies. Five-minute headways were maintained in the afternoon. Unfortunately, this heavy traffic caused two cars to burn out one motor each, a problem which would occur with increasing frequency in the months ahead.

Early Days on the Line. No sooner had service begun between Scranton and Pittston in May 1903 than the L&WV became involved in a labor dispute. Organizers from the Brotherhood of Railroad Trainmen and from the Amalgamated Order of Street Railway Employees vied with each other for the right to represent the workers. The Laurel Line management felt it was caught in the middle in this dispute and attempted to remain neutral. Conn said the company had no objection to its employees joining any union they wished. This position was not satisfactory to John H. Devine, president of the Scranton Central Labor Union, which was affiliated with the Street Railway Employees Union. When Devine failed to convince the management, he announced a boycott of the Laurel Line. This apparently had little effect, since the trains ran as scheduled and passenger loads increased steadily during the year.

Throughout the summer, conferences were held between the Laurel Line officials and the labor organizers. Only four of the employees were members of the Brotherhood of Railroad Trainmen, since that union required a year's experience in railroading to be eligible. As most employees were new to the railway field, they opted for the Amalgamated organization. By October the situation was resolved when most of the men paid their initiation fees and became members of the Street Railway Employees Union. A charter was granted for a local chapter, after which the Central Labor Union lifted its boycott. Ironically, the workers soon shifted their allegiance to the trainmen's union after the L&WV management assumed control. The peaceful resolution of this conflict was the harbinger of future labor–management relations, which remained harmonious until the post-World War II period.

From the beginning, the third rail presented hazards. In August of 1903, Chester P. Wilson, superintendent, wrote to Conn that "Evilly disposed persons are continuing to charge the wire fence separating our right-of-way from that of the Erie between Market Street and Pine Street, Pittston, by placing wires, barrel hoops, etc., in contact with the third rail." In Scranton, the line had to build fences behind the company houses near the terminal to protect the children from falling down the bank to the third rail. By September, two people, the first of many, had been killed by carelessly touching the highly charged power rail. Part of the problem was that the rail was open without a wooden cover as used on other lines and even though the right-of-way was fenced, some passersby managed to wander onto the tracks. The *Wilkes-Barre Record,* in an editorial, demanded more protection, ". . . for the territory through which the road passes abounds in ignorant people who cannot be made to understand their own peril."

Shortly before service opened to Wilkes-Barre, the company placed ads in community newspapers warning the public of the danger from moving cars and the third rail. Despite these and other notices, the third rail was to prove dangerous throughout the life of the line.

In its first months, the line attracted a growing number of passengers; 38,830 people were carried in May 1903. In September this had grown to 86,594 and to 133,786 in December. During January 1904, the last month the line was operated by the Transit Contract Company, 153,568 fares were collected. This growing passenger load taxed the equipment of the company. The managers had initially hoped to run trains of two cars and one trailer, but the motors were inadequate for this task because they were only 150 horsepower and it was estimated that at least 200 horsepower would be necessary. An even greater difficulty was the failure of the management to order any trailers in the first place. Very shortly an order for new motor cars would be placed, which would alleviate the equipment shortage. In the meantime, the single units took about 40 minutes to cover the almost 20 miles between Scranton and Wilkes-Barre.

Inevitably, accidents and problems began to appear. No sooner had the line opened in May 1903 than the first accident, a relatively minor one, occurred. The *Scranton Times* reported that "Contractor Muldoon's horse was killed by the third rail after jumping over an iron fence on Moosic Street and pulling the carriage down the embankment." Muldoon and his driver were slightly injured by being thrown out of the vehicle. The first worker to be killed was a night watchman employed by contractor John R. Lee. Evidently, he had been riding on a northbound ballast train near Connell Crossover and either fell or was knocked off by the third rail warning sign, striking his head as he fell. For the more fortunate employees who survived accidents, the company had an agreement with the Moses Taylor Hospital in Scranton to care for injured persons for the sum of seven dollars per week.

Winter weather disrupted operations during several snow and sleet storms. The third rail scrapers failed to work, causing considerable flashing, delays and unfavorable comments from the public. A power failure on February 1, 1904, compelled the line to lease steam locomotives and cars from the Lackawanna Railroad all day until repairs could be made. The breakup of ice on the Susquehanna River and subsequent flooding caused the southbound track to collapse near Wilkes-Barre. More than 50 carloads of ash and stone were needed to repair the damage.

On the same day of February 1904 the contractors relinquished control and the Lackawanna and Wyoming Valley Railroad formally assumed responsibility for the operation. Up to this time, a total of 822,789 passengers had been carried, producing a gross revenue of $120,000, all of which was kept by the contractors to help meet construction expenses. To that date, a total of $5,961,000 had been spent to build the line and much work remained to be done, notably the one-mile tunnel under Scranton. Potential traffic exceeded the line's ability to carry it and it was decided to order 16 additional cars equipped for multiple-unit control. Management hoped that freight and express traffic would really develop after the completion of the Scranton tunnel. Although expenses were excessive for a small operation, optimism prevailed. George Smith told his board of directors that ". . . the interests of the companies are in the hands of experienced and competent men who will work with great energy and intelligence to secure the desired results. . . ."

Events would soon indicate that the managers would need all the "energy and intelligence" they could command in order to make this heavily overcapitalized line earn a profit.

A southbound combine is about to enter the Laurel Line's most expensive engineering project, the tunnel under Crown Avenue in South Scranton.

Edward S. Miller

2

A New Line in the Valley: Electricity Challenges Steam

ON FEBRUARY 1, 1904, the Lackawanna and Wyoming Railroad formally took control of the new line. Many challenges confronted it. It was becoming clear that the high cost of construction and the related interest payments on the bonds would be a severe financial strain on the company. The minimum gross earnings necessary to carry the debt and pay expenses were estimated at $500,000 per year, but at this point the line was earning approximately $250,000 per year.

It was hoped that the planned Scranton tunnel, which would be an additional financial burden, would attract freight traffic, thereby increasing revenue. But George Smith, president of the L&WV and vice president of Westinghouse's Security Investment Company, refused to predict that the line could appreciably increase its gross earnings, contenting himself with the statement that the readers of his report should draw their own conclusions. It was apparent that only a combination of the most fortuitous circumstances and skillful management would see the Laurel Line grow and prosper.

A factor which contributed to the gloomy financial outlook was the top-heavy organizational structure of the L&WV. Heading the company were President Smith and Vice President Conn. Under the latter were departments of transportation, traffic, treasury, accounting, power, roadway and construction, each with its manager and assistants. On the operating level there were a trainmaster, inspector, chief engineer, master mechanic, roadmaster, chief dispatcher, dispatchers, electric line foremen and even railway policemen, to say nothing of station agents, baggage masters, night officers, janitors, power station employees, car repairmen, clerks and, of course, motormen, conductors and track workers. Complicating this bureaucratic maze was the interrelationship with the Westinghouse organization in New York City. Certain officials, such as President Smith and Treasurer Carl M. Vail, worked for Westinghouse in New York and were rarely on the scene. This complex structure not only obstructed the decision-making process but it also created a financial drain on the firm's resources and proved ineffective in responding to the major financial and operating challenges confronting the company.

Parks Increase Ridership. In common with electric railways throughout the country, the Laurel Line management looked to the development of parks as a means of attracting riders. Several possibilities for park sites existed along the route of the company. One of the most promising and, as it turned out, the most enduring was Rocky Glen near Moosic, five miles south of the Scranton terminal, owned by the Rocky Glen Water Company. An artificial lake was available, which could be the center of an attractive park when properly developed. As early as August 1, 1903, a temporary station had been established there.

That winter, a local promoter, Arthur Frothingham, who controlled the water company, placed ads in the area newspapers, stating, "The skating at Rocky Glen is fine, going down, take the Laurel Line." As the 1904 season approached, the L&WV signed a 10-year lease for a consideration of one dollar with the Rocky Glen Water Company for a tract of land. A permanent station was constructed, consisting of sidings for 12 cars, platforms, an overhead bridge and large station building. An arrangement was made to sell combination railway and park admission tickets in the Laurel Line stations. Prices ranged from 25¢ from Scranton to 40¢ from Hancock and 50¢ from Wilkes-Barre. The improvements were successful and more than 200,000 people were carried to and from the resort in 1904.

Another site developed along somewhat different lines was Valley View Park, located at Inkerman, about one mile south of Pittston, and 13 miles south of Scranton. The company secured a tract of 20 acres of woodland under a 10-year lease at $150 a year. A siding was constructed and a pavilion and booths erected. The company conceived Valley View Park as a sedate picnic area for church groups and civic organizations. Rules were published in May 1905 stating that "Intoxicating liquors in any form will not be allowed on the premises . . . and the use of profane and indecent language or disorderly conduct will not be permitted." Evidently these strictures created a pleasing environment, since schools, churches and private clubs made extensive use of the facilities. More than 75,000 people were carried to and from the park in that season. However, the company was anxious that Valley View

VIEW IN ROCKY GLEN PARK, SCRANTON, PA.

Rocky Glen, Summer resort, near Scranton and Wilkesbarre, Pa.

Best regards to Mr Myers from Florence Kennedy

SWISS COTTAGE AND PARK, ROCKY GLEN, NEAR SCRANTON, PA. 6299

not compete with the more exciting entertainment provided at Rocky Glen.

Apparently, eating picnic lunches and playing simple games were not enough to attract sufficiently large crowds to Valley View Park. In March 1905, Conn urged that a baseball field and stands be erected there for use by a new Wilkes-Barre team that was entering the New York State League. He estimated it would cost $2,500 to lease additional land and build the field, but that the company would earn the money back in the first year from the fares of fans attending the games. Perhaps the team was not too good; in any event, Smith was unimpressed and no action was taken. Valley View remained essentially a picnic grove, closed on Sundays, and it gradually declined in importance as changing tastes and times led people to seek more exciting diversions.

A third site considered ripe for development was Linwood Park on the Dunmore branch, about one and one-half miles northeast of the Scranton terminal. Frederick Ingersoll, who had prior experience in developing amusement parks, approached Conn in 1905 and suggested constructing a modern amusement park in the Nay Aug section of the city of Scranton. He pointed out that Valley View was too far from Scranton to operate profitably even if it were converted to an amusement park. In contrast, Linwood Park would have the advantage of being within the Scranton city limits and with its alluring mechanical attractions might generate more revenue by attracting riders from Pittston and Wilkes-Barre as well.

Smith was receptive to the idea. He agreed the L&WV would furnish power to the park if Ingersoll would provide the necessary transformers and electrical connections. Conn was charged with handling the details but was told not to make them public. An informal agreement was made with Ingersoll, whereby he would pay his power bill "from time to time." Smith expected the facility, now called Luna Park, to attract 150,000 people annually and thus supply much-needed revenue.

On February 24, 1906, newspaper readers in the Scranton, Wilkes-Barre area were treated to an announcement from the Scranton Luna Park Company proclaiming that "The greatest popular amusement event in the history of northeastern Pennsylvania will take place May 21 when the . . . Park Company will throw open the superb Luna Park. . . ." The broadside went on to describe the park in glowing terms as having "unexcelled facilities for summer outings and picnic parties." It promised that the park would present ". . . a dazzling picture of quaint and massive architecture," illuminated at night with 50,000 electric bulbs. The picnic grove would be ". . . quaint bridges, shady paths and quiet nooks." There would be a large casino building housing a dining room and numerous attractions such as Shoot-the-Chutes, Trip to Rockaway, the Mystic River and a score of others.

A month after this bombastic announcement, on March 15, 1906, the Scranton Luna Park Company was formally organized. Major Thomas Penman was elected president. A lawyer, he was then collector of internal revenue for the Twelfth District of Pennsylvania, and he had been an active promoter of the companies which then made up the L&WV Railroad, still serving on its board of directors. However, the real force behind the company was Frederick Ingersoll.

The L&WV's involvement with the Scranton Luna Park Company is an example of the managerial incompetence which plagued the line during these years. Extensive advertising announced the park would open on May 21, 1906, and contracts were made with various groups. Because of delays, the opening was postponed until May 28 and an immediate loss of $800 was incurred because of the forfeiture of a contract with a band booked for May 21.

In order to survive, amusement parks have always needed a turnover of rides and other attractions. These postcards of Rocky Glen Park, mailed between 1907 and 1910, bear little resemblance to the park in 1952. Conspicuously absent in the early years was a roller coaster; the park had several over the years.

Author's Collection

Postcard images are the only survivors of Luna Park. Although the Laurel Line could find nothing to attach in 1912, the park remained open until 1916. The cards reproduced on this page were all postmarked in 1907 and 1908.
Fred W. Schneider, III Collection

EDISONIA, TEMPLE OF MYSTERY, SCENITORIUM, MERRY-GO-ROUND; LUNA PARK, SCRANTON, PA. 1001

Even worse was a misdirected involvement with the Scranton Railway Company, operators of the local streetcars in the city. Although the facility was served by the Laurel Line's Dunmore branch, the park company expected that most of its patrons would arrive on the cheaper and more frequent city cars on the Nay Aug line. This route ended about one-third mile from the entrance to Luna Park and the promoters expected the railway company to extend its line to the gates. However, it failed to do so and this meant that passengers from the streetcars had to walk to Luna ". . . through the woods over uneven, unlighted ground, strewn with boulders, and undergrowth." Approximately $800 had to be spent to build and light a road from the Nay Aug picnic grounds to the Luna Park entrance.

In sharp contrast to the indifferent policy followed by the Scranton Railway Company, the Laurel Line spent in excess of $7,000 in building a station, a siding, platforms and a subway under the tracks, even though the bulk of the park patrons preferred the more convenient streetcars. The park company estimated that Scranton Railway increased its receipts by at least $20,000, although it would not even build a shelter for its passengers.

Frederick Ingersoll had obviously overestimated patronage and profits. Penman criticized his plan as "extravagant and too expensive." Ingersoll estimated there would be 500,000 paid admissions, whereas only 325,000 people attended. In 1907, patronage declined further and the park company faced a severe financial crisis. Ignoring this abysmal situation, the flamboyant Ingersoll had the audacity to write to Smith in September asking him for a favorable letter ". . . stating what Luna Park has done for your lines in the way of . . . increasing traffic." He planned to incorporate this in a catalog showing his amusement park enterprises.

Although Smith was naively willing to comply, Conn informed him the Luna Park Company still owed the L&WV more than $1,100 for its share in the construction of the subway under the tracks. By January 1908, Ingersoll was forced into involuntary bankruptcy by a Pittsburgh court, after the failure of another of his enterprises. The L&WV managed to secure a judgment against him for $1,100 in 1909, but it could not find any assets against which to make an execution. Finally, in 1912, the L&WV gave up trying to collect from Ingersoll and wrote off the sum as a bad debt. In the meantime, Luna Park continued to decline. The Laurel Line realized very little return from its considerable investment in Luna Park and the facilities to serve it.

The only remaining park was Rocky Glen and it, too, was beginning to experience a decline in attendance. Late in 1906, Arthur Frothingham offered to sell the entire property of 120 acres, including the water company, to the L&WV for $250,000, saying that other groups were interested in acquiring the site. Conn interpreted that to mean the Scranton Railway Company or the Delaware and Hudson might be working to purchase the property. Smith, when informed of this, replied that the difficult financial condition of the road precluded purchase. He felt certain the Laurel Line would carry the bulk of the patrons to the park regardless of who owned it.

Conn initiated a new scheme to encourage patronage to the park in February 1910. The Scranton baseball team with a schedule of 20 games was looking for a sponsor. Conn estimated the team would draw about 5,000 people per game, producing $25,000 for the season in gate receipts. Twenty percent of this would accrue to the L&WV to pay off its projected investment of $10,000. Additionally, about 80,000 15¢ fares and 20,000 40¢ fares might be collected from fans attending the games.

Smith was unimpressed. "I am of the opinion that it would be unwise for the company to make the investment at this time," he wrote Conn. "Nevertheless, I would encourage the location of a ball park in every way possible, other than investment of money." By the end of the 1913 season, Rocky Glen temporarily ceased to operate as an amusement park and remained open only for swimming and ice skating through 1918.

The Scranton Tunnel. Far more important to the Laurel Line than amusement parks was the need for a tunnel under the south side of Scranton. From the first, the promoters recognized the necessity for bypassing the formidable 4 percent grade over the Scranton Hill. Only in this way could freight traffic be effectively developed. Initially, the company ran surveys for a double-track bore to be lined with porcelain-faced brick. Since this was too expensive, a single-track tunnel became a necessity.

Following its usual practice, the L&WV incorporated still another company to finance this venture. On July 7, 1904, the L&WV Rapid Transit Company signed an agreement with the *Lackawanna Tunnel Company* under which the transit company would build a tunnel from Elm Street under Irving Avenue to Front Street, a distance of almost a mile. As finally constructed, however, the tunnel would follow an alternate route under Crown Avenue. The tunnel company would pay the transit company $1,249,000 in cash, stock and first and second mortgage bonds. The transit company, in turn, signed an agreement on July 29 with the Security Investment Company, under which the latter would actually construct the tunnel. The L&WV would pay one million dollars for the work.

Construction began on July 5, 1904, by sinking two perpendicular shafts from Crown Avenue to the proposed roadbed. By September 8, in addition, headings were extended in both directions from each shaft, making a total of six headings. During construction an average of five carloads daily of excavated material was taken from the tunnel and added to the rip-rapping along the Susquehanna River right-of-way. A continuous opening from portal to portal was completed by mid-July 1905.

A number of problems had to be solved in the course of construction. Many property owners above the route of the tunnel filed claims for damages. The company's attorney advised offering them $50 for each vacant lot and $200 for each house lot. Conn rejected this proposal, saying that it was unlikely the property owners could definitely pinpoint the spot under their property where the tunnel passed and that they would be unable to secure an injunction stopping the work. As a result, the work proceeded uninterrupted, but eventually the L&WV had to settle the damage claims on an individual basis. Several people who had homes above the tunnel complained that their wells had suddenly gone dry. Naturally, they blamed the blasting in the tunnel. Later, they too had to be compensated.

An unexpected difficulty occurred when the tunnel was nearing completion. Fire was discovered in an upper vein of a coal deposit within the northern section of the bore. A gang of 25 men was employed daily for several weeks fighting this fire. In order to contain the conflagration, the contractor spent nearly $1,800 uncovering lower veins in the deposit. Since the tunnel was being lined with timber at the time, Conn wrote to George Francis, a Westinghouse engineer, suggesting that metal sheathing be added to the timber to prevent sparks from locomotives from setting it on fire. Francis pointed out that the fire risk would be greater with sheet iron than without it. The iron would give the sparks a place to lodge, he said, whereas they would harmlessly fall from the exposed timber after striking it.

When completed, the tunnel was 4,750 feet in length; 1,300 feet of it was solid rock, 750 feet lined with masonry and 2,700 feet lined with timber. At its narrowest point, it was 17 feet wide and 22 feet high. Each of the two shafts was 10 feet by 20 feet, the southerly shaft being 110 feet in depth and the northerly shaft 180 feet. The maximum grade was one percent.

The tunnel zone extended from Scranton Junction in the north to Connell Junction on the south, a distance of 2.2 miles. It consisted of double track from each junction to the immediate vicinity of the nearest portal, where it converged into a gauntlet track laid through the tunnel proper. Initially, due to a shortage of 90-pound rail, the northbound track in the tunnel was laid with 75-pound steel. Later, this was replaced with the 90-pound type. A total of 120 incandescent lamps, wired six in a series, illuminated the tunnel.

As protection for the trains on the gauntlet track, an electric staff system, consisting of a staff box located in an operator's booth at each junction, was installed. Each box had 40 staffs. In time, some of the staffs became bent or damaged, but as there were so many of them, it made little difference. In 1915, a canvas bag was placed at South Scranton so conductors of southbound trains could throw the staffs into it, to be retrieved by the operator and returned to the machine. Trains were ordered not to exceed 25 mph when passing this point.

To facilitate train movement at Connell Junction, the company bought a mechanical interlocking plant with a 10-lever Stevens machine which controlled the two junction switches and the movable crossing frogs. Switches at Scranton Junction were rearranged and a 1,500-foot siding was laid next to the main track to permit switching at the carhouse without interfering with scheduled trains. A subway was constructed under the tracks between the carhouse and the power plant to permit employees to go safely from one point to the other. Regular service began through the tunnel on October 19, 1905, and by formal agreement on January 31, 1907, the L&WV accepted the tunnel as fully completed.

A new timetable was issued, which showed 81 trains operating on the main line. Of these, 13 limited trains operated each way through the tunnel, as did seven southbound and four northbound locals in the afternoon. Three minutes were saved. The remaining locals used the old "over-the-hill" route. Dunmore cars ran on a 20-minute headway, turning left at Laurel Junction at the top of the Scranton hill.

Contractor's equipment sits before the bridge over Roaring Brook at the north portal of the Scranton tunnel in the summer of 1905. The L&WV's over-the-hill line ascends the embankment above the tunnel mouth. Note the slope opening of a small coal mine, which was unearthed by the excavation in the lower right.

Charles Sobeck, Edward S. Miller Collection

Workers pause (ABOVE) to gaze at the camera man at the north portal of the Crown Avenue tunnel. The chute on the right was used to funnel materials from the existing trackage above. *Charles Sobeck, Edward S. Miller Collection* **Early in October, 1905, photographer Horgan mounted his 8x10 view camera on a tripod and exposed the glass plate from which this print of the newly completed South Portal (RIGHT) of the Scranton tunnel was produced. Note the gauntlet track which required two third-rails. This was simplified during the first half of the 1920's to a single track and one less conductor rail.**
Edward S. Miller Collection

The timber-lined section of the South Scranton tunnel, deep under Crown Avenue, presented photographer Horgan with a very difficult lighting problem. Flash powder to illuminate the bore (bulbs were not available in 1905) was apparently ignited behind the support columns for several hundred feet. If one of the assistants goofed and got himself or his shadow in the picture, there was no problem because photo studios of that period all employed expert photo retouchers.
Edward S. Miller Collection

Concrete lining gives way to natural rock (LEFT) near the south end of the tunnel. *George Arents Research Library, Syracuse University* **The single-track bridge (RIGHT) over Roaring Brook was erected in 1905 as part of the tunnel project. Idle rails remain in place today, over the bridge and through the tunnel.**
Edward S. Miller Collection

New Cars Arrive. While the promoters had grandiose plans for the construction of a high-speed, heavy-duty railroad, they were seriously deficient in their original order for equipment. Not only was the number of passenger cars inadequate, but their single-ended construction and the absence of MU controls seriously hampered their use. Consequently, on January 25, 1904, an order was placed with Westinghouse, Church, Kerr for 16 additional wooden cars, 13 of which were to be straight coaches and three to be combines. All would have dual controls with MU capabilities and train doors. Westinghouse gave the order for the carbodies to the John Stephenson Car Company of Elizabeth, New Jersey.

The motors, Westinghouse type 50 F with a gear ratio of 18-59 for each car, and trucks were shipped separately along with the brake equipment for assembling in the Scranton car shop. The passenger cars were numbered 11 to 23 and the combines 111 to 113. Control was by Westinghouse's turret type unit switch group. Each car cost $13,000. Some of the cars were ready for service by mid-June 1904, and the remainder by August, except for numbers 11 and 12, which collided early in the month. The L&WV designated them as type M equipment.

As with the original cars, the new equipment was outfitted with Ohmer fare registers, which were rented from the Ohmer Company for a cost of $2.04 per day for the total of 23 registers. At Conn's insistence, no advertising was carried in the cars.

Both the new and old cars developed problems. During the first winter of operations, many passengers complained about the cold during the early morning runs. Some threatened to go back to the Delaware and Hudson trains if things did not improve. The problem was alleviated by fixing defective heaters or by the obvious step of turning on the heaters before the car was placed in service.

An even more serious situation with the type M cars involved defects in the brake rigging. Because of failures on the type L cars, Superintendent Chester Wilson was concerned about the adequacy of the braking system. In October 1903, before the new cars were constructed, Conn, at the insistence of Wilson, urged that separate brake shoes and brake rigging for the hand brakes be installed, since, in the event the air brakes and rigging failed, it would still be possible to stop the car safely. Westinghouse, Church, Kerr engineers rejected his suggestion, claiming there was insufficient room on the motor truck and that putting the hand brake on the trailer truck would produce about 40 percent efficiency. With considerable reluctance, Conn accepted the argument of the Westinghouse engineers.

As the vice president feared, problems developed. Late in June 1904 the brake rod on one of the new type M cars broke near the Nay Aug station on the Dunmore line. Fortunately, the car was not on a severe grade at the time.

Conn's apprehensions were realized in 1905. On May 5, a passenger car from Wilkes-Barre experienced a broken brake rod while descending the 4 percent grade from Laurel Junction to the terminal. As the car rapidly gained speed, soon reaching 40 miles an hour or more, motorman Cohen reversed the motors, thereby turning them into electric brakes. This effectively cut the speed to about 10 miles an hour as the car swung into the terminal loop. Sitting directly ahead of it was a Dunmore local car, which was preparing to leave. Inside the runaway, pandemonium reigned. One of the passengers and the conductor panicked and jumped from the end platform to the station floor, suffering severe bruises as a result. Another passenger, referred to as "an unknown Italian," also jumped, but he was carried away by his friends before anyone could learn of his injuries or his identity. The remaining eight passengers suffered minor injuries and no one on the Dunmore car was hurt. Damage to the runaway was minor, but the rear platform of the Dunmore car was crushed.

In his report to President Smith, Conn could find no reason for the failure of the brake rod. He commended the crew, who stayed with their train and followed the emergency procedures outlined in the rule book for stopping the car with the electric motors. The crew of the Dunmore local was taken to task, perhaps unfairly, for failure to anticipate the runaway and move their car out of its path.

One outcome of the accident was that Westinghouse executives finally became concerned about the brake equipment. A representative from the Baldwin Locomotive Works was sent to Scranton to see if separate shoes and rigging for a hand brake could be applied to the cars, as Conn had originally proposed. While this investigation was underway, still another accident struck the line on July 10, which fortunately was minor. On one of the type L cars a fire suddenly broke out. It was only with difficulty that the motorman was able to stop the car before the wooden body burst into flame as the fire had damaged the airbrake equipment. Once more, Superintendent Wilson pressured Conn to install a hand brake on the rear platform, and a circuit breaker as well, to stop the car in case the motorman should be incapacitated. After considerable investigation, the type L cars were modified in the manner indicated at a cost of about $180 each. In the meantime, similar equipment was added to the type M cars.

It is interesting to note that the argument which most impressed the Westinghouse executives in New York was Conn's statement that in the event of a serious accident, the company would be liable for heavy damages, since the neighboring Wilkes-Barre and Hazleton Railroad cars were equipped with the separate shoes and rigging for the hand brakes.

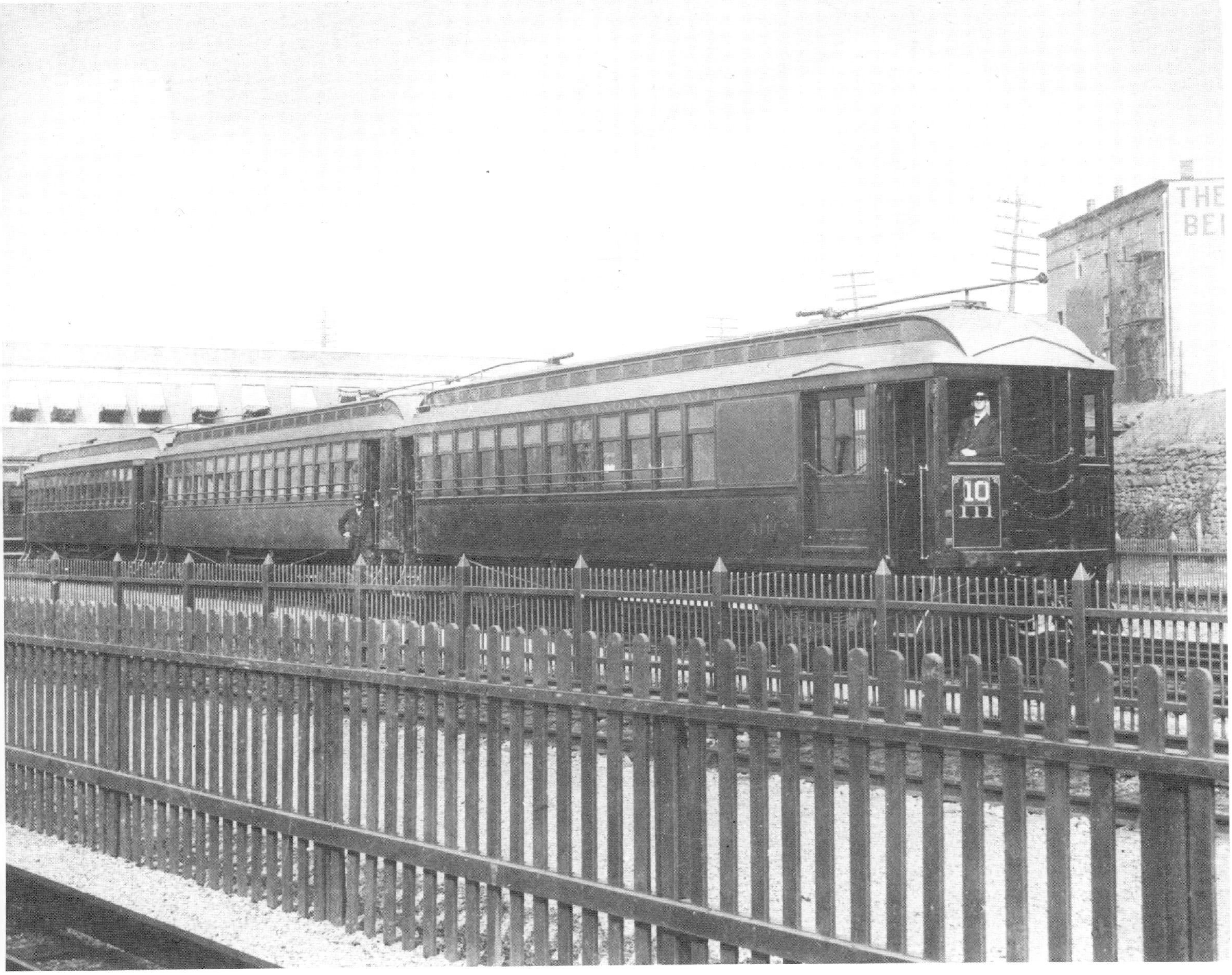

Three of the type M cars, led by car 111, the "Powhatan," were posed at Scranton soon after they arrived on the property.
G. Dewey Shrader, Edward S. Miller Collection

Electric Locomotives. The completion of the Scranton tunnel in October 1905 and the consequent prospects of developing a significant freight traffic prompted the company to order electric locomotives as the express motors would be inadequate for the task. As it turned out, the Westinghouse Electric and Manufacturing Company had on hand a machine built in 1895 as an experimental model. It weighed 84,000 pounds, had two four-wheel trucks and four 150-hp motors. Built as an AC/DC machine, it had undergone extensive testing on the Pennsylvania Railroad. The engine was now surplus and Westinghouse saw an excellent chance to sell it to the Laurel Line.

President Smith in December 1905 ordered Westinghouse to send an engineer to Scranton to confer with Conn about the details of the purchase. The engine, referred to as number 6 by Westinghouse, had broken sills and would need new motors, control equipment, transformers, cab flooring, and other repairs. The engineers estimated it could attain a speed of 45 miles per hour when running light and approximately 15 miles per hour when hauling full loads with a drawbar pull of 13,000 pounds. With a 33 percent overload on the motors, the speed would be 10 miles per hour and the drawbar pull 27,000 pounds. The locomotive was 32 feet in overall length, nine feet wide, had a distance between truck centers of only 15 feet and had 42-inch-diameter wheels.

Conn accepted the Westinghouse offer, except that he wanted a bow trolley which the motorman could change himself and thus eliminate the need for a brakeman while switching. Since there was only a small section of overhead trolley line on the L&WV, this was not done and the locomotive arrived with a pantograph, which was soon replaced by a conventional spring-base trolley pole.

After extensive correspondence and many modifications, the locomotive was delivered on April 15, 1906, and was numbered 401. With it came a bill for $17,500, which was the first firm figure the L&WV officials had received. When placed in service, the modified engine was considerably heavier, weighing 97,000 pounds, had a solid steel floor and two very heavy four-wheel trucks. It was equipped with multiple-unit control, MCB couplers, automatic air brakes sufficiently powerful to handle 15 cars, and was designed to pull eight 100,000-pound loads at 25 miles per hour on level track.

Because of anticipated increases in freight traffic, the L&WV ordered a duplicate of engine 401, which was subsequently built by Westinghouse and delivered in 1907 as number 402. It was similar in most respects, except that it was slightly heavier, weighing 106,000 pounds. Oddly enough, Westinghouse charged the same sum for number 402 as it did for the older, experimental locomotive.

A somewhat faded 401 heads a train of gondolas in the Scranton yards about 1910. As usual, the crew has stopped work while the camera man is in the vicinity. The motorman's uniform with coat and tie was rather unusual for his calling and was not changed to the more traditional overalls on the Laurel Line until 1914.

G. Dewey Shrader, Edward S. Miller Collection

Cows, Cabs and Free Passes. Ludicrous as it seems, in its first years of operation the company was confronted with an invasion of cows. In one week during October 1904, seven of the bovines were killed with a resulting outcry from farmers.

To protect the livestock, officials stationed watchmen at the public crossings where the beasts wandered onto the right-of-way. They were instructed to make a pretense of doing work, but their real function was to goad the animals off the property in the hope that they would not return.

The L&WV made an exclusive agreement with the Scranton Transfer Company to provide cab and carriage service at the terminal station. This firm was permitted to solicit passengers, use the L&WV telephones and could call upon the baggage master for assistance. In return for this monopolistic privilege, Scranton Transfer promised to have a carriage and two cabs available throughout the day.

What was to become a perennial issue for the Laurel Line management arose even before service started. This was the bedeviling question of free passes. The first known request for transportation came from the Wilkes-Barre Water Company in 1902, which wanted passes for its employees. In a response, which was the exact opposite of his actual practice, Conn wrote that he ". . . personally [was] strongly opposed to the granting of free transportation for any purposes whatever. . . ." Despite this seemingly hostile attitude, only six months after service began, the line was carrying from 100 to 300 free passengers daily. They included employees, workers at Rocky Glen, local railroad officials and newspapermen.

Perhaps one reason for Conn's change of attitude was that he had written to the presidents of at least 40 railroads offering to exchange passes. They ranged from the mighty Pennsylvania Railroad through the nearby Lackawanna, Erie, Lehigh Valley, Jersey Central, Delaware and Hudson, and New York Ontario and Western, to such remote carriers as the Rutland, Grand Trunk, Kansas City Southern, Texas and Pacific, Louisville and Nashville, Southern, Pere Marquette and even New York City's Interborough Rapid Transit Company. He was not always successful. For example, a vice president of the New Haven acknowledged his request with this smug reply: "When the New Haven connects with your little old road and operates it under a lease, I shall probably use [your pass]."

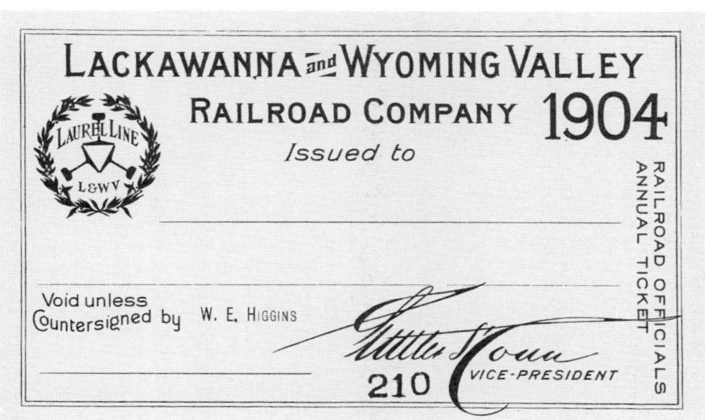

Graced by the distinctive signature of Charles F. Conn, this pass typifies those issued in the early years. *Edward S. Miller Collection*

Local politicians soon discovered that they could procure passes and the company was showered with requests from various officials. Evidently, those who had influence, either real or potential, received them, while others, who were judged by management to be less useful, were denied. For example, Scranton city councilmen, members of the sheriff's department, Lackawanna county commissioners and even Jury Commissioner P.J. Kelly received passes in 1905. In the latter case, Conn's attorney advised giving one because, "I can foresee that there might arise occasions where a jury commissioner could be of service and I think it advisable to comply with Mr. Kelly's request."

Apparently, the lowly deputy inspector of factories for Bradford County was judged to be less influential, since Conn baldly responded to his letter by writing, "As we do not issue complimentary passes, I am not in a position to comply with your request." A Republican city commissioner of Wilkes-Barre was told, "It is not our custom to grant free transportation to political organizations." By 1906, reason was beginning to prevail as President Smith instructed Conn to limit his pass requests to railroads operating in the area.

Car Failures and Modifications. The equipment continued to present challenges to the Laurel Line's management. In 1906, it was decided to rebuild the original cars, designated type L, to a double-ended configuration with MU controls to make them compatible with the newer type M cars. As built, these cars had a motorman's cab on one end, a platform on the other, hand controls and straight air brakes. As a result, they could not operate in trains and, considering the absence of an automatic block signal system, service was slow and spotty during periods of heavy travel. In the rebuilding, vestibules and train doors were added to both ends of the cars, as well as multiple-unit controls, automatic air brakes and Van Dorn couplers.

The high-voltage wires were enclosed in grounded conduits under the carbody and were sheathed in asbestos, thus minimizing the danger of short circuits and electrical fires. Nonetheless, chemical fire extinguishers were placed in each car and the crew members trained in their use. Express cars 201-204 had their cabs rebuilt with a glass window permitting a view to the rear and an interior sliding door. This enabled crewmen to have a comfortable station during freight operations. Typically, Conn underestimated the cost of this work. In 1906, he wrote that an outlay of $40,000 would be necessary to modify the equipment. The final figure, as reported in 1910, was more than $80,000.

In September 1906, before it could be rebuilt, car 10 was destroyed by fire. The company decided to order a new body from the John Stephenson Company in Elizabeth, New Jersey, which closely resembled the newer type M model. The new car had passenger and smoking compartments, transverse seats and mahogany finish. Its overall length was 51 feet 2 inches, with width over sills 8 feet 10 inches. Retaining its original number, 10, and the name *Capoose,* the car arrived in Scranton in January 1908 and entered service late in April.

The newer type M cars developed a serious problem by 1908. It appeared that the ends of these wooden cars were sagging and that the floors were deformed. Superintendent Wilson launched an extensive investigation to determine the nature of the trouble. He concluded that the bolsters of the

Flanked by two newer Stephenson-built type M cars is rebuilt type L car number 5. The rebuilding of the type L cars with multiple-unit or MU control allowed any car in the fleet to run at the head end of a train of up to three cars, with the motorman in the lead car simultaneously operating the motors and brakes on all cars in the train. Capacity of the line was increased because two or three cars could now be run with the same spacing between trains as was previously possible with single cars. The type designations L and M indicated that the cars had either L-2 drum controllers or unit-switch type multiple-unit control, respectively. Technically, after 1908, the L&WV had only type M cars, a fact that was indicated by special instructions in the timetables for use by employees. However, shop personnel continued to use the type L and M designations long after there were no type L cars on the property. Although precise dating of this illustration at Rocky Glen Park is impossible, it is probable that number 5 was the first of the rebuilt cars and that the picture was taken in April or early May, 1907. *Dr. Edward Steers Collection*

cars, which were designed by Westinghouse, Church, Kerr, were not of sufficient thickness, and that the three-quarter-inch bolts on each end snapped under the pressure of constant service. The side sills remained in proper place, but the center sills had sprung with the result that the car floor was higher at each king pin than it should have been.

After consultation with the chief engineer of the J.G. Brill Company (the Stephenson Works were temporarily closed), the line decided to order new cast steel bolsters of a design submitted by the Atha Steel Casting Company of Newark, New Jersey. The total estimate for 32 bolsters was $1,525 and it cost the company an additional $2,038 for labor and material to install them. About 10 days' work involving at least eight men was required to apply the new bolsters to each car. The cars were modified one at a time over a period of six months in 1909. Furthermore, the multiple-unit controls were rebuilt to eliminate many petty problems, which had caused frequent delays and interruptions in train service.

Cold weather caused moisture to form and freeze in the air pressure line. On the cars with automatic control, this resulted in malfunctions which led to breakdowns on the road, since the unit switches were activated with compressed air. The trouble was blamed on the absence of a slide reducing valve which Conn maintained had been specified in the original orders for the new controls. The Westinghouse Electric and Manufacturing Company argued that such valves were not necessary and recommended draining the air tanks frequently in cold weather and that a new grade of packing be applied to the valves to eliminate leaking. After extensive correspondence on this matter, Conn's views prevailed and the new valves were installed.

Another difficulty was excessive arcing from the third rail shoes during snow and sleet. This was corrected by installing new steel spring seats for supporting the third rail beams on the trucks. Surprisingly, the line managed to operate reasonably well during snowstorms without a plow. The only equipment purchased for snow removal, aside from the usual scrapers attached to the third rail shoes, was a Q&C Priest Snow Flanger obtained in July 1905 for $750. It was attached to work car 301 and used when necessary.

Employees complained about cold drafts and inadequate heat in the motormen's cabs. A workers' committee, which met with the company officers, blamed the ineffective fastenings of the end doors, loose trap doors and large holes in the switchboard where the cables came through for the excessive drafts in the cabs. The master mechanic admitted the trap doors fit poorly and charged the builders with sloppy construction. Bolts were used to secure the end doors and the switchboards were corked to stop the flow of air.

More serious were failures of motors which caused cars to break down while in service. Westinghouse Electric claimed the motors were designed for 500-volt current rather than the 650 volts used, which makes one wonder why they sold them to the L&WV in the first place. They recommended the purchase of new, larger motors of at least 200 horsepower. The L&WV was not about to bear this major expense, but other steps were taken to minimize the interruptions. For example, the current from the powerhouse and substations, which had been fed to the third rail directly outside of those buildings, was now carried on the overhead feeder wires to points a half-mile distant from either side. This helped to produce a lower voltage on the line and reduced circuit breaker openings.

All cars were withdrawn from service in Scranton after one round trip and given a layover of 30 to 50 minutes to allow the motors to cool. Making a virtue out of necessity, Conn noted that this practice would enable the company to clean the interior of the cars and keep a proper temperature in them in the winter. A special bulletin was issued to the motormen reminding them to keep the power off at all breaks in the third rail and on all long downgrades. These measures helped to reduce the instances of motor failure, but maintenance costs still remained high.

Many minor improvements were made. Among them was the slotting of the commutators in the car motors to permit the use of softer brushes, which extended the life of the average brush from only 500 to 8,000 miles. Experiments were conducted with different makes of trolley bases, and new springs and stands on all equipment eliminated broken or bent trolley poles. Three new light circuits were added to the passenger cars and new type brake shoes were placed on all the equipment. Work car 301 was enclosed on one end and equipped with wrecking tools and an extra air compressor.

One improvement which was not made, despite pressure from the Department of Public Health, was to install drinking water on the cars. Conn said water was available in the stations at Scranton, Pittston, Hancock and Wilkes-Barre, and this was sufficient.

As the L&WV now had two electric locomotives, steam locomotive number 10 was offered for sale in April of 1907. Despite its poor condition, Conn confidently quoted a price of $2,000, even though the American Locomotive Company told him it would cost $750 to put it into working order. Although many dealers were contacted, none was interested. Finally, in December 1909, John B. Watson, an equipment dealer, found a buyer for $1,000, the Central Iron and Coal Company of Lebanon, Pennsylvania. Much relieved, Conn ordered the locomotive shipped via the DL&W and either the Pennsylvania or Philadelphia and Reading lines. But the engine never got out of the L&WV yard in Scranton, since these carriers refused to accept it because of its poor condition.

Central Iron complained about the delay in delivery and finally canceled its order. Thoroughly frustrated and determined to get rid of this troublesome machine at all costs, Conn offered it to Luria Brothers and Company of Reading for scrap, receiving $500 for it. It was loaded on flatcars and shipped at last on February 2, 1911.

As freight business picked up, Conn decided the company needed more cars. Originally, the line received 10 30-ton boxcars numbered 500-509 from the American Car and Foundry Company's Berwick, Pennsylvania, plant. One of these, 507, was destroyed in a wreck in 1909 and replaced by a duplicate car with the same number at a cost of $1,000. Also among the early equipment were two flatcars, 1001-1002, and two gondolas, 1003-1004. Conn tried to persuade President Smith to buy new boxcars and 10 gondolas from McGuire Cummings Manufacturing Company in 1906. He told Smith the interline freight business was growing slowly and he argued the Laurel Line could earn money from per diem charges on foreign roads. According to his figures, the L&WV paid $865 over a three-month period for the use of foreign cars, while at the same time it received $66 from the steam roads for their use of L&WV cars. Smith refused to approve buying new cars but he authorized Conn to look for cheaper, used cars.

In the same year Conn bought 10 hopper-bottomed gondola cars, apparently from the DL&W. With a capacity of 50,000 pounds, they were painted brown with L&WV lettered on the sides and numbered 701-710. In 1907, 10 additional cars, numbered 711-720, were purchased from the DL&W for $345 each. Two years later, Conn secured five flat gondolas for $2,500, numbered 601-605, to replace the 1000-series cars. Number 1001 was dismantled in 1917 and its trucks used in the shops. Of the three remaining cars, 1002 was assigned to the railway department, 1003 was used for ash handling, and 1004 was assigned to the electric line department.

All told, it seems the L&WV spent an excessive amount for repairs to passenger cars and renewals of electrical equipment. For the period from 1905 to 1910, the company expended $95,500 and $79,700 for these expenses respectively. Considering the age of the cars, this total expenditure of $175,200 appears to be unreasonable. This does not include the $80,000 cost of converting the type L cars to double-ended MU cars.

One factor which undoubtedly contributed to these high repair costs was shoddy maintenance procedures at the Scranton shop. T.R. Langon, an inspector for the Westinghouse Electric and Manufacturing Company, visited the property in February 1912 and wrote a report highly critical of the L&WV shop forces. Despite the huge outlay for repairs and maintenance, Langon said that the motors were troubled with ". . . grounded fields, grounded armatures, short-circuited fields and . . . armatures, severe flashing . . . and poor brush holders." He attributed this to the "rundown condition" which they had been in for three years. He found the suspension system holding the motors to the trucks was so badly worn that the motors almost hit the underside of the car and caused wear on the bearings and axle collars. The control apparatus was in ". . . a most dilapidated condition, with every indication of absolute negligence on the part of the inspection department." The condition of the automatic turret type control he found as ". . . simply rotten." Langon reported that during a recent cold spell, delays ranged from 20 minutes to more than two hours and that "mighty few trains, if any," left on time because of control and motor failures.

The inspector was alarmed because equipment superintendents on neighboring lines, such as the Lehigh Valley Transit Company and the Wilkes-Barre and Hazleton Railroad Company, were well aware of the difficulties the Laurel Line was having. He criticized the recently appointed master mechanic who lacked experience and knowledge in both the electrical and mechanical fields. Langon felt this situation was very detrimental to Westinghouse, pointing out that ". . . when other properties are in the market for equipment . . . our competition immediately refer them to this road, which is called by all . . . 'The supposed Westinghouse road,' and which experiences the most equipment trouble of any line in the area.

Although it is possible Langon was exaggerating the condition of the Laurel Line equipment, since Westinghouse had already given up on the company, it is more likely that his charges were valid. Certainly Conn's local management had to bear some measure of responsibility for these conditions. Perhaps his lack of experience in the electric railway industry prior to 1903 contributed to the situation. It may be significant that when Westinghouse sold its interest in the L&WV in 1913 Conn did not remain with the Westinghouse corporate family. Instead, he secured a position in the cement industry.

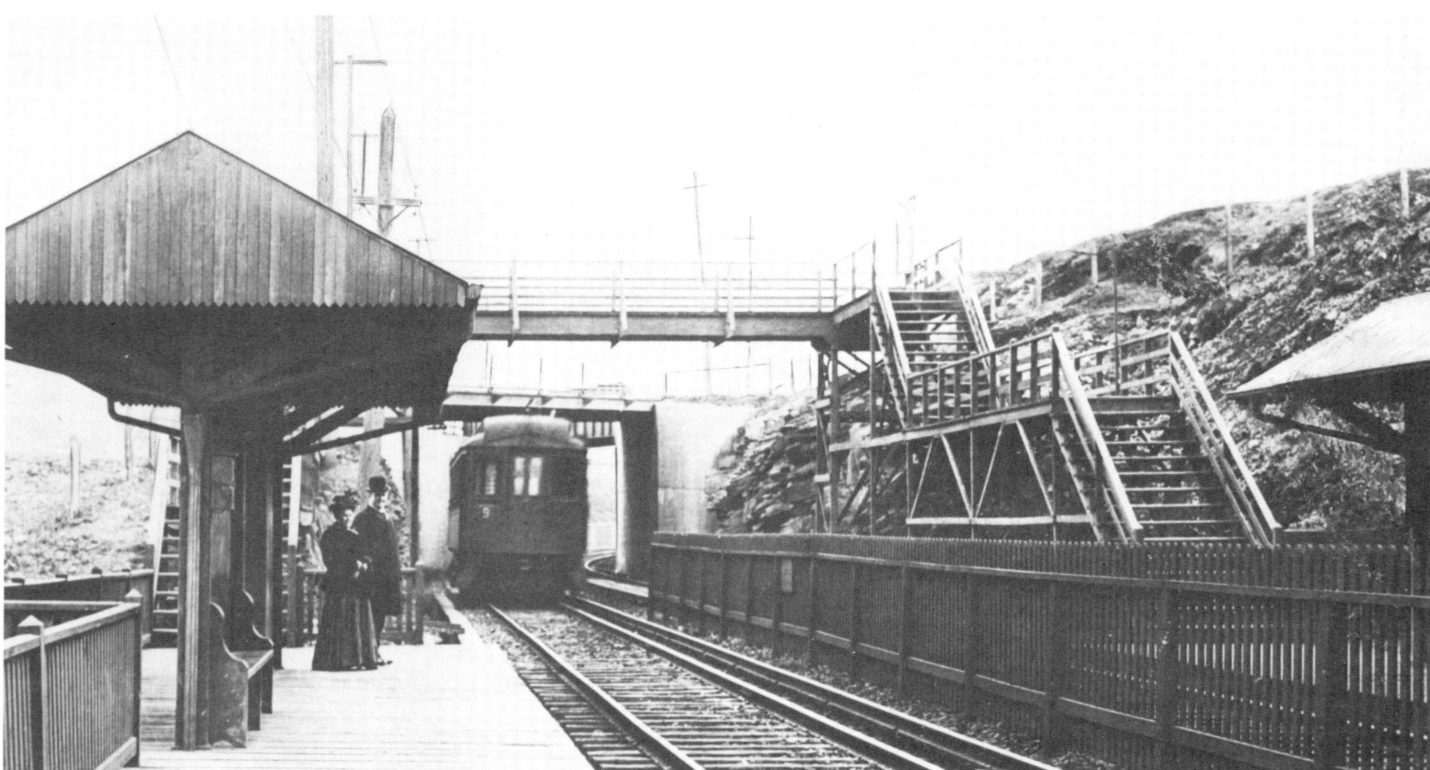

Perhaps this couple was among those inconvenienced by the frequent delays experienced by the L&WV. The locale is the South Pittston station at Plank Street in 1906. The arriving train is headed north.

Edward A. Brady, Edward S. Miller Collection

Despite continuing maintenance problems, two type M cars race southward at Blanchard's Farm, a short distance below Inkerman station, in 1908. Nearby was a siding used for loading tomatoes from the farm.
Edgar J. Boland, Edward S. Miller Collection

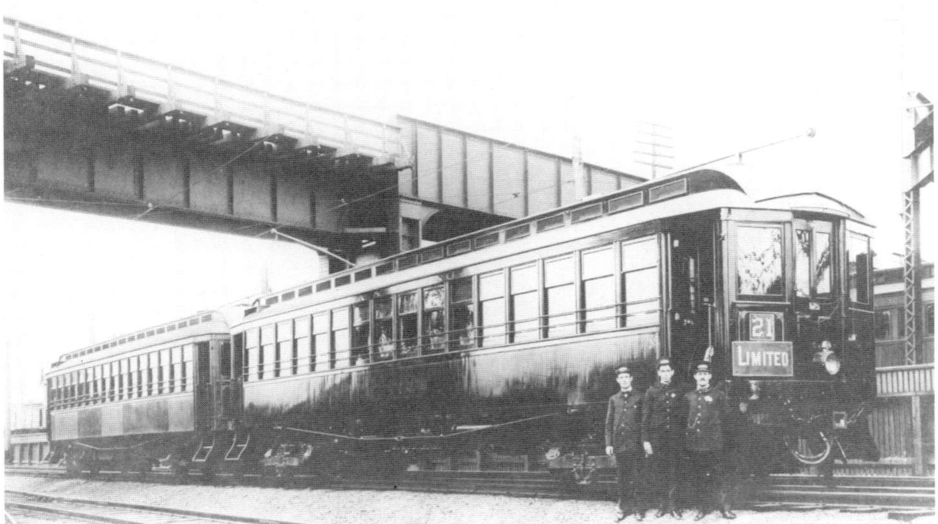

Nattily attired in their L&WV uniforms, the young crew of limited train 21, left to right: Peter Manley, C.J. Newcomb, and Jesse D. Caryl, stand beside car 6, named the "Onas." The second car is number 3, the "Nescopeck." Both were type L cars recently rebuilt to operate with the newer type M cars. Attention is directed to the mismatched paint on the lower body panels of car 3, indicative that after only six years of service selected rotten wood boards had been replaced, a testimonial to the high cost of maintenance of wooden cars. The pair was photographed under the Wilkes-Barre & Hazleton viaduct in Wilkes-Barre.
Jesse D. Caryl, Edward S. Miller Collection

L&WV Power Company. Taking a page from its own past, in 1906 the L&WV chartered yet another corporation. Unlike the earlier ones, this new entity, the *Lackawanna and Wyoming Valley Power Company,* was not formed simply as a device to raise capital. Its main purpose was to generate and sell electricity at cost to the railroad, and for a profit to mines and industrial firms in the Scranton area. The availability of electric power might serve as a magnet, which would attract manufacturing concerns to lands along the right-of-way, thereby increasing freight traffic. To meet this demand, a 2,500-kilowatt turbogenerator and two additional 400-horsepower water-tube boilers were added to the facilities in the powerhouse at a cost of $103,000. The generator, built by Westinghouse, was designed to deliver three-phase alternating current at 25 cycles and 2,200 volts. Also installed were three 1,000-kilowatt, 25-cycle, 2,200 to 22,000-volt transformers with accompanying switchboard, wiring and pumps.

Originally the power station boilers burned number one buckwheat anthracite coal, which cost about $1.75 per ton. Management realized, by checking with other power companies in the area, the boilers could be modified to burn much finer and cheaper sizes of coal or even culm, which was a mixture of fine coal and refuse remaining after the breakers extracted the better grades of fuel. Barley coal, for example, sold for just 50¢ a ton and only a slightly greater quantity was needed to produce the same amount of steam. Therefore, the Roney Mechanical Stokers were replaced with Neemes Shaking Grates. The estimated savings were $1,000 per month. The ash siding adjoining the powerhouse was electrified, in part by overhead trolley and the remainder by third rail, to permit the electric motors to remove the residue in standard freight cars.

By 1907, the power plant was operating at maximum capacity which made it impossible to perform regular maintenance. As a result, efficiency decreased and the cost of operation increased. By 1908, repairs could be postponed no longer and an extensive overhaul of the equipment was made. Full efficiency was not restored since some of the machinery had deteriorated beyond the point of repair. At the same time, the average cost for a ton of fuel jumped from 50¢ to 75¢ and the wages of the workers also rose. The net result was a 48 percent increase in the cost of operating the plant.

The sale of commercial power, increased passenger service and the growing freight traffic compelled the company to improve its distribution system. A new substation was built at Avoca in 1907, consisting of a one-story brick and concrete building with a three-story tower next to the freight siding. Under the building were mines operated by the Delaware and Hudson Coal Company. The L&WV had to purchase certain pillars of coal in the top vein and then fill the openings between them with culm for a distance of 25 feet around the outside of the building.

This structure at Hancock housed a converter substation and a combined passenger and freight station. The facility opened in 1903 when service commenced south of Pittston and was renamed Plains in 1911. The substation took 22,000 volts alternating current from the transmission line and produced 650 volts direct current to operate trains on the south end of the railroad. The A.C. source was first passed through transformers to step it down to 390 volts. This low-voltage A.C. powered a pair of Westinghouse 40 KW rotary converters, i.e. 650 volt D.C. generators driven by three-phase A.C. induction motors. Initially, one of the two converters was held in reserve. Increased traffic forced the L&WV to add a 50 KW General Electric synchronous converter. The 1940-vintage photo shows the step-down transformers and all three converters.

Interior by Harold Machler, Exterior from Edward S. Miller Collection

Even with these precautions, mining engineers would give no guarantee that the substation would not be affected by subsidences in the mines. In the building were two 500-kilowatt rotary converters and three 375-kilowatt transformers for the purpose of changing 22,000-volt, three-phase AC into 650 DC. There was room for the possible future installation of three additional transformers and one more converter. As placed in operation, the Avoca substation increased the third rail voltage and made it more even on the system as a whole. This helped the car motors and made it easier for trains to maintain their schedules.

To supply power to the Cheesman Chemical Company and several other industries in the area, a small substation was constructed at Virginia. The brick building was 20x20 feet and contained three 75-kilowatt, 25-cycle, 22,000 to 2,200-volt transformers. A one-mile transmission line was extended from the substation to Connell Junction to serve industries at that point. It was placed in service in November 1906.

The main power station was improved by the installation of a 1,500-kilowatt rotary converter, capable of transforming 390-volt, three-phase AC into 650-volt DC. This new machine not only increased the potential output of the power plant, but it also made it possible to shut down some of the other equipment for necessary repairs.

It was fortunate that this additional machinery had been installed, since a year later serious difficulties developed with one of the Corliss steam engines. Because Superintendent Chester Wilson, who had previously supervised operations, resigned, Conn assumed direct control of the powerhouse in January 1909. Reports came to Conn of reduced power and upon investigation it was discovered the turbine blades were so badly corroded that it was necessary to replace a large number of them. Conn blamed wet steam for this condition and also for problems with the pistons and valve stems.

Westinghouse engineers made the necessary repairs, which included replacing 21 rows of turbine blades. Their bill of more than $800 precipitated a lengthy dispute, since Conn maintained he had asked for only 19 rows of blades to be replaced. He also felt the engineers had made errors and were charging the L&WV for the additional time they needed to correct them. Mr. C.V. Holmes, of the Westinghouse Machine Company, not only dismissed Conn's charges but he found that because of an error the bill was actually $62 less than it should have been. In a nit-picking manner which became typical of the road, Conn kept the issue alive by asking for further consideration of the matter. The somewhat exasperated Holmes ended the correspondence by stating the bill was reasonable and fair and he had no desire ". . . to enter into a prolonged discussion of this account, as was the case with the last charge we had against you for repairs. . . ." Reluctantly, Conn approved payment.

An interesting sidelight at this time was a study conducted by the Delaware, Lackawanna and Western Railroad concern-

Added four years after the line opened, the Avoca converter station made the third-rail voltage more uniform on the northern part of the railroad. Originally it housed a pair of Westinghouse 50 KW induction motor-driven rotary converters. A similar 40 KW unit was later added. In addition to the power plant in Scranton and the converter stations in Avoca and Plains, the power distribution system included alternating current substations at Virginia, Rocky Glen, Ewen, and No. 3 Shaft for commercial customers. *BOTH: Edward S. Miller Collection*

ing possible electrification of its tracks in the Scranton area. Evidently, the steam road was weighing the merits of electric traction for pusher service. It asked Conn whether the L&WV could supply current for two or three locomotives. Smith was cautious since negotiations were underway with another carrier, probably the Delaware and Hudson, for the same purpose. This, combined with commitments to area collieries, would extend the power plant beyond its capacity. As it turned out, the DL&W dropped the matter.

By 1912, the plant was operating at a reasonable level of efficiency. Lightning caused occasional problems, as in June and July 1911 when it badly damaged a turbogenerator. Power was sold to the L&WV railroad at cost; therefore, as expenses rose or fell, so did the charges to the railroad company. In 1911, the power company earned over $102,000, of which more than $86,000 was received from the railroad and $16,000 from other companies. Included in the latter were the Pennsylvania Coal Company's Ewen breaker, the Wilkes-Barre Colliery Company, the Carleton Coal Company, the D&H Coal Company and the Scranton Electric Company, which purchased 3,000 kilowatts of energy. In 1911, the power company paid two dividends of 10 percent, which went directly to the L&WV.

The Conn Administration. The managers of the L&WV realized the line had to generate more traffic if it were to meet its operating expenses and cover its bonded indebtedness. In December 1906, Conn tried to persuade the Lehigh Valley Railroad to run a through car, possibly a sleeper, from Scranton to New York City. The L&WV would haul the car from Scranton to its connection with the Lehigh Valley in Wilkes-Barre; Conn argued that there would be profitable traffic for this service. The Lehigh Valley refused to make such an arrangement, saying its terminal was too congested and the service would not pay. It is more likely the Lehigh Valley management realized that most Scrantonians going to New York City would prefer the shorter and more direct route of the DL&W.

A few years later, Conn tried another tactic. He attempted to persuade the Lehigh Valley to permit the Laurel Line to sell sleeper space on the Wilkes-Barre to New York City run, from the Scranton ticket office. Charles Lee, general passenger agent of the Lehigh Valley, was interested until the Delaware and Hudson learned of the plan and expressed ". . . quite serious objections." Lee said the Lehigh Valley could not afford to antagonize the D&H, since they had a large freight and passenger interchange with it.

The D&H probably was sensitive to any dealings with the L&WV, as the electric line was a brisk competitor with the steam carrier for local and through traffic between Scranton and Wilkes-Barre. Studies conducted by the L&WV in 1906, 1907 and 1908 showed the electric road was carrying approximately 85 percent of all fares along the route. Although some of the Laurel Line's trade may have been won at the expense of the D&H, it created an essentially new market, since its service was faster and cheaper than the steam roads.

A further analysis of L&WV traffic revealed some interesting profiles of its passengers. About 29 percent of the fares rode between the terminal cities, providing 50 percent of the revenue; the 15 percent who journeyed between Pittston and Scranton provided 16 percent of income; riding between Pittston and Wilkes-Barre was 13 percent of the people who paid, surprisingly, 13 percent of the revenue; and the remaining 21 percent of the income was derived from the 43 percent who rode among local stations. Aside from an overall population growth in the area, it was unlikely the Laurel Line could substantially increase the number of passengers carried. As it turned out, the new automobile and competing bus were to prove far more serious a challenge than the steam roads.

Occasionally there were unusual incidents which increased the passenger trade. For example, on March 13, 1904, the company handled 10,000 riders, many of whom were traveling to points along the Susquehanna River south of Pittston, to view the huge blocks of ice and the resulting conditions caused by the spring thaw. In the afternoon, 10-minute service was provided from Scranton and two extra cars were put on in an effort to clear up the crowds at Pittston and other stations. Even with these measures, some individuals had to wait as long as an hour to board a train.

A variety of special movements generated revenues. Among them were many excursions to Rocky Glen and Valley View, school picnics, band concerts, race meets, dances, church outings, Shrine meetings and funeral parties. The company charged 50¢ for every dead body transported on the line. Each corpse had to be accompanied by an adult who paid the regular fare. There were frequent specials required for various theatrical companies, who performed such excit-

Charles F. Conn carried out policy formulated by the Westinghouse empire during his tenure with the Laurel Line.

No doubt, car number 8, seen at Wilkes-Barre in 1905, was among the equipment pressed into service to accommodate the throngs who flocked to the L&WV in March, 1904, to view the ice-choked Susquehanna River. The motorman, on the right, was Robert Francis Boland, who helped to build the Laurel Line, and lived until 1955.
Edgar J. Boland, Edward S. Miller Collection

ing melodramas as *Kidnaped in New York* and *The Great Automobile Mystery*. Such movements demanded not only a passenger coach but also an express car to carry the baggage and scenery.

Sometimes large conventions produced heavy traffic, such as those held by the Elks in Scranton and the Knights of Malta in Wilkes-Barre in May of 1908. The company estimated it earned more than $1,700 from the delegates to these two gatherings. A Redmen's state convention in June brought 2,000 delegates to Scranton, who ". . . proved to be good travelers and were continually riding over the line." Conservatively estimated, increased earnings amounted to $1,000.

Traffic could fall as well as increase, as the summer of 1909 revealed. Not only was the weather unfavorable, with frequent rain and showers, but the miners were working only three days a week due to the large supply of coal on hand caused by a mild winter and stockpiling in anticipation of a strike. New resorts north of Scranton at places such as Lake Winola, Brookside, Moosic Lake and Lake Ladore were becoming very popular, thus reducing travel to parks along the Laurel Line. The same was true with new resorts on lines serving the city of Wilkes-Barre such as Sans Souci Park, Nuangola, Harvey's Lake, Lake Carey and Mountain Park. In addition, the steam railroads were running excursions to places such as the North Jersey Coast, Lake George, the Adirondacks, Niagara Falls, Atlantic City and New York City. At the same time, travel to resorts along the Laurel Line fell.

With business declining, Conn decided to employ a detective agency to determine if the conductors were reporting all their fares. It was noted that some conductors were registering all their fares in bunches instead of recording them at the time of payment. Many minor errors were discovered, such as failing to report one or two fares. One conductor was described as being asleep in the last seat and others spent their time talking to the motormen or passengers. These individuals were reprimanded and warned against repetition.

As the line entered its fourth year of operation, an impressive number of trains was run. The Dunmore branch, which was 2.48 miles long, ran 110 local trains daily, each taking six minutes, serving three intermediate stations and averaging 24.8 miles per hour. On the main line, 136 local trains were scheduled every day via the "over-the-hill" route through South Scranton. The cars served 15 intermediate stations between the terminal cities, which were 19.2 miles apart. Speeds averaged 28.1 miles per hour for the running time of 41 minutes. There were also 26 limited trains daily, which ran through the tunnel, thereby shortening the distance to 18.44 miles. These trains served two intermediate stations, were scheduled for a running time of 30 minutes and averaged 36.88 miles per hour. The last half mile between River Street and the Wilkes-Barre terminal had several grade crossings which slowed the trains. Average speeds between Scranton and River Street in Wilkes-Barre, therefore, were even higher and more indicative of the true speeds.

Photographer Horgan produced this excellent series of photographs of the South Scranton station for evidence in the McCracken case about 1907. Although the tunnel had been opened in 1905, most mainline trains still climbed over the hill. For example, the June, 1909, timetable listed 63 mainline trains over-the-hill in each direction Monday through Saturday. The tunnel was used on workdays by only 13 southbound and 12 northbound limiteds plus 4 southbound and 5 northbound scheduled freight trains. On August 23, 1914, as a result of the runaway earlier that summer on the hill, all mainline trains commenced operation through the tunnel. This station, then renamed Maple Street, became the southern end of the three-cornered Dunmore line. In spite of the temporary nature of the line, service was provided to this stop for 42 years. Today, a foot bridge crosses Interstate highway 81 where Maple Street once crossed the tracks. *Dr. Edward Steers Collection*

Cars on the "over-the-hill" route averaged 29.32 miles per hour, while cars via the tunnel route averaged 39.58 miles per hour. In addition to the many passenger trains, the company operated eight regular freight trains and two regular Adams Express trains between Scranton and Wilkes-Barre daily through the tunnel. The locomotives hauled extra freight trains during the early morning hours as needed.

To its stockholders and board members, the management liked to portray the Laurel Line as an efficient and well-run organization. Smith claimed the management was popular with the local communities and he noted the line was commonly regarded as "The Peoples Railroad." This may have been the case, but the "Daily Report of Train Service" forms show delays were all too frequent. Included in the litany of reasons were, ". . . freights working on the track, poor power, changing jumper cables, burned-out motors, fuses blowing, air pump failure, loading newspapers, mine cave-ins, broken trolleys, hot journals, bad rail, no sand, broken trolley wire, and cows on the track."

Less serious but still vexing were occasional minor problems. Conn became concerned about the unsanitary practice of some passengers spitting while on the cars. He wrote to the state board of health in Harrisburg and received a typical bureaucratic response. The health officers said they would deputize employees to enforce the ordinance against expectorating and even give them the power of arrest. There is no evidence Conn was willing to risk altercations between train crews and offending passengers. No doubt peaceful persuasion was used.

A particularly irate patron, W.G. Harding, secretary of the Pennsylvania Central Brewing Company of Scranton, wrote to Conn on January 30, 1905, complaining the temperature in car 113 registered at 55 degrees when he boarded it in Wilkes-Barre for the express run to Scranton. According to Harding, the electric heaters were cold and therefore the car was defective. Before Conn could respond, he received a letter the next day on the same subject. This time Harding reported the temperature was only 40 degrees, and he sarcastically suggested that some loose bricks stored at Wilkes-Barre should be heated by the agent on his stove and used to warm the passengers' feet. Conn replied that he was investigating.

Unsavory characters periodically frightened passengers along the line. One man called to say that his daughter had been accosted by ". . . a hard looking citizen with red hair and corduroy pants" on the platform at Nay Aug Park Station. He told her he was "fixing the wires." The approach of a train frightened him off and no harm was done. Miscreants were dealt with harshly by the management of the line when they were caught. One individual who made the mistake of falling asleep in boxcar 504 was hauled to the magistrate and fined $5.00 and $1.50 in court costs. When he said he had no money, he was sentenced to 15 days in the county jail.

A trivial problem, but one which consumed much time in correspondence and which would recur frequently, was stealing from sanitary napkin machines in the terminal stations. The original machine in the Wilkes-Barre station, which was a new model designed to discourage theft, was itself stolen in 1910. Its replacement was an older model, much easier to break into. Therefore, the thieves contented themselves with merely taking the contents and the money rather than the machine.

When Conn asked for replacement napkins for those stations, the manager of the National Case and Carton Company told him the company ". . . had more trouble with the various cases placed in Wilkes-Barre and Scranton than in all other cities combined owing to the rough element in these places." He rather impractically suggested that the women's toilet be locked at night after the matron left. Conn chose to ignore this alternative and cheerfully pointed out that conditions had so improved in Scranton that the profits there would offset the losses incurred in the Wilkes-Barre station.

Advertising. Oddly enough, for a company in need of as much revenue as it could get, the Laurel Line refused to carry advertising in its cars. Several firms wrote to Conn seeking a contract to place advertising, but the vice president responded that the company did not wish to use these materials in its cars. Occasional ads were placed in local newspapers listing service and schedule changes, but no concerted campaign was undertaken to attract passengers. At one point, Superintendent Chester Wilson suggested the company distribute free newspapers to its morning patrons, but Conn rejected this idea. Another employee proposed the company launch a campaign through pamphlets, mailing lists, prizes and streetcar cards to entice people to live along its lines. He felt the attractions of suburban living should be stressed to woo potential future riders.

Again, nothing happened. The only significant improvement was the erection of an electric sign in front of the Wilkes-Barre station in 1909. Despite its low-keyed approach to advertising, the *Wilkes-Barre Record,* at least, was impressed with the system. It wrote a glowing editorial in March 1912 praising the progressive management of the company and chiding those pessimists who had been doubtful of its success. In reality, the company was already in financial difficulty and reorganization was on the immediate horizon.

Despite its reluctance to carry advertising in the cars, management was proud to erect this spectacular electric sign on Lackawanna Avenue at Cedar Avenue in Scranton during 1908.

George Arents Research Library, Syracuse University

When it came to soliciting freight, the L&WV was more aggressive in advertising. Early in 1906, Conn had an attractive industrial folder prepared, with a romanticized color cover depicting the valley as the residents would no doubt like to see it. This brochure contained several pictures of the company's facilities and a detailed map of its route. The text urged industries to consider the merits of locating their plants along the Laurel Line.

Given wide circulation through the local boards of trade, the folder was well received. The Wilkes-Barre freight agent of the Lehigh Valley Railroad described it as the neatest, most concise, and ". . . withal wholly creditable publication of its kind I have ever seen. The map is a particularly good one." Despite the effort, the solicitation of freight traffic was, on the whole, disappointing.

Freight Grows Slowly. One major difficulty in developing freight traffic was the reluctance of the steam railroads to permit the L&WV to become a part of their tariff arrangements. In 1904, Col. William Mattes, a consulting engineer, approached the vice president of the Delaware and Hudson about the possibility of making an interline agreement. The D&H official was not receptive to the proposal, claiming that while his company had nothing against the L&WV as a railroad, they ". . . have strongly objected to the attitude of Mr. Westinghouse in the matter, he having had for many years important contacts with the various steam railroads and now is entering the field as their competition." This attitude and the general reluctance of the steam railroads of the day to recognize the new upstart electric lines would present the L&WV with an uphill struggle to gain parity with its competition.

Nonetheless, the Laurel Line persisted in its efforts and arguably achieved modest success. Joint tariffs went into effect in February 1905 with the Erie Railroad and with the Lehigh Valley in March. On July 1, 1906, agreement was reached with the DL&W. Conn solicited the aid of the local boards of trade in his efforts to secure additional interchange arrangements. These organizations, however, were of little help.

A dream long held by Conn and other officials was to make the L&WV part of a through route between the west and New England. Although there was only a remote possibility of the large carriers considering such a plan, Conn pursued it with a tenaciousness that revealed his frustrations and perhaps a growing desperation. For example, in 1905, when the New Haven Railroad was developing its interchange yard at Maybrook, New York, Conn hoped to persuade the Pennsylvania to route its New England traffic to Wilkes-Barre, then over the Laurel Line to Scranton, picking up a proposed connection with the New York, Ontario and Western, at that time controlled by the New Haven. The O&W would carry the traffic to its interchange with the New Haven at Campbell Hall.

The Pennsylvania, at the time, was routing some of its freight through Philadelphia and Trenton and on to the Lehigh and Hudson River Railroad, which also connected with the New Haven. Other freight carried by the Pennsylvania Railroad was floated across New York harbor to the New Haven facilities on the Harlem River. As was true with many of Conn's proposals, this did not reach fruition. In fact, the Pennsylvania Railroad refused to make any interline agreement with the L&WV.

Nor was Conn any more persuasive with the New York Central. In December 1906, B.F. Wyly, the traffic manager, met with his counterpart on the Vanderbilt road. He presented a strong case for an interline agreement between the two companies. Wyly stressed the steam railroad charter and characteristics of the L&WV and said it was capable of handling the very heaviest of freight traffic. After claiming the Laurel Line was managed by men "possessing thorough steam railroad experience," he asserted that the L&WV was ". . . one of the highest types of railroad construction, equipment and speed; and in the march of progress is somewhat in advance of the most substantial and best equipped railroads in the country." After evaluating these exaggerated statements, or perhaps because of them, the New York Central refused to make an arrangement with the Laurel Line.

Undiscouraged by this rebuff, Conn again approached the Pennsylvania, this time through the good offices of the DL&W. John Crawford, the general freight agent of the Lackawanna, interceded for the electric line with the officials in Broad Street Station. The mighty PRR refused to budge and would not enter into any division or through rates with the L&WV. Crawford asked whether it was because the Laurel Line used electricity and suggested the small road might complain to the Interstate Commerce Commission. The cagey PRR man would say only that he could not make an exception in favor of the L&WV and that Crawford was personally familiar with their position.

Conn began to take a more realistic view of the freight situation by 1909. In his yearly report to Smith, he noted the large investment in freight facilities had ". . . thus far been comparatively unremunerative." The vice president correctly indicated the inaccuracy of the projections made by the original promoters. The communities served by the Laurel Line were major producers of anthracite coal and little else. Nearly 90 percent of the coal mined was owned and carried by parent railroad companies. Therefore, the principal movement of other freight was inbound and most of this was also handled by the steam railroads. Conn felt the L&WV must encourage manufacturers to locate along its lines and to seek through traffic with other railroads.

In analyzing the competitive situation, he noted the Delaware and Hudson and Jersey Central Railroad basically paralleled the L&WV and therefore were not likely to make traffic arrangements. The Erie, likewise, reached most important points with its own rails. The Lehigh Valley sent most of its business destined for Scranton over the D&H, while the DL&W did find the Laurel Line useful for serving Wilkes-Barre.

Having made this dispassionate analysis, Conn once more painted an idealistic scenario for projected growth. Aware that some freight for Boston and other points in New England moved from Wilkes-Barre to Albany over the Delaware and Hudson, he argued the L&WV should build a 3½-mile connecting track from its Dunmore terminal to the New York, Ontario and Western at Throop, north of Scranton; construct an expanded yard at Wilkes-Barre and purchase a few more locomotives. He was not discouraged by the fact that the projected Throop extension would have a two percent grade and a number of sharp curves.

Conn felt the New Haven, which controlled the O&W, would be receptive to the idea, since it would get a greater share of the revenue. The Laurel Line would benefit by serving as a bridge route between the PRR in Wilkes-Barre and the O&W near Scranton. He projected a minimum of 100 cars daily over this route.

Conn's thinking was faulty in three respects. First, he overestimated the likelihood of the PRR agreeing to such an arrangement and ignored its still open hostility to any kind of cooperation with the L&WV. Second, he assumed the New Haven would make considerable expenditures to improve the O&W's Scranton division. Third, he neglected the possibility of developing coal traffic from independent mines near the right-of-way.

By 1909, the L&WV had been successful in making tariff arrangements with a number of lines and freight associations. Included among them were the DL&W, Lehigh Valley, Erie, Lehigh and New England, the Southeastern Freight Association through the Cincinnati gateway for lumber, cast-iron pipe, oil and petroleum, and the Central Freight Association, which included most midwestern carriers. The line was also a party to east and westbound tariffs on the Great Lakes and indirectly reached southern cities via the Lackawanna Railroad to New York, in connection with the Old Dominion Steamship Company, the Clyde Line, the Mallory Line and the Morgan Line. Significantly, no agreements had been made with the Baltimore and Ohio, Pennsylvania, New York Central, Western Maryland or any of the transcontinental lines. Obviously, the interchange traffic handled by the L&WV was carried only a short distance and therefore produced minimum revenues.

During this period, total cars handled per month ranged from 371 in September 1906 through 1,478 in June 1910. Approximately five to eight percent of the total carloads were non-revenue cars in company service. The heaviest volume of freight was in the form of inbound commodities for various local industries. Coal was the most important item produced in the area and shipped out. In terms of volume, these figures would seem to be fairly impressive, but the L&WV received such a small share of the revenue split, because of its short distance, that it is doubtful it made any significant profit.

All these efforts to increase freight traffic nevertheless achieved a certain degree of success. A number of companies located their factories along the right-of-way and, collectively, they provided a fair amount of business for the electric line. Most of these were located in the Meadow Brook Valley, just south of Scranton. They included the Warren-Ehert Roofing Company, which produced about 25 tons of crushed slag per day, and the Connell Brick Company, which had an output sufficient to fill 400 cars per year shipped from a siding 200 feet long trailing from the northbound track. The Cheeseman Chemical Company, formerly of Alexandria, Virginia, constructed a plant in Moosic. To serve this plant, the Laurel Line installed a 575-foot siding from the northbound track. Operating from a tract of land eight acres in size, the company had a traffic of about 40 tons per day. Eventually this firm failed, and it was reorganized in 1911 as the Barium Products Company and continued to supply some revenue to the Laurel Line.

The L&WV became directly involved with the Meadow Brook Crushed Stone Company. John R. Lee had installed a stone crusher to provide ballast for the right-of-way during the period of construction. Since a demand for the product of the crusher existed, and the Laurel Line gained freight revenue from shipping it, the Security Investment Company purchased a controlling interest in what was then called the Meadow Brook Crushed Stone Company. About 25,000 tons of crushed stone were shipped in 1906 and 1907, earning about $4,000 for the railroad.

Unfortunately, the plant was inefficient, and it soon became indebted to the L&WV. Rather than lose this investment and traffic, Conn proposed that new electric machinery be installed at a cost of almost $15,000. This was done in 1908 and the crusher reopened. The entire capital stock of $40,000 was owned by the L&WV. Because the market was soft, production fell from 25,000 tons in 1908 to 20,000 tons the following year. In 1911, the plant was leased to the MacDonald Construction Company for one year, for a minimal rental of $3,000. Eventually, it failed entirely and thus another of Conn's plans had turned sour.

The Scranton station was the bureaucratic center of the railroad. Housed in the second floor of the 83′-6″ x 46″-0″ building were the offices of the general manager and auditor on the street side, and the superintendent, paymaster and engineering staff on the track side. Curiously, the cab stand in the left portico had no entrance door into the station. Office expansion would alter the Cedar Avenue side of the structure during 1923-1924.

Edward S. Miller Collection

Express car 201 is being loaded at the Wilkes-Barre freight house. Faintly visible on the North Pennsylvania Avenue side of the structure is a horse drawn wagon, the prevailing means of draying freight prior to 1910.
Edward S. Miller Collection

Located in Pittston was the American Slagolite Company, which manufactured concrete blocks. It had an output of three cars per week. In Wilkes-Barre were several industries, including the Crocker Grocery Company, Whitman Stationery Company and Hoover Mercantile Company, all of which had warehouses located on the sidings in the city. Several other firms announced plans to build factories at various points, but these did not materialize.

To serve these and other plants, the company built a frame freight station, 48 feet by 28 feet, including warehouse, office and platform on the west side of the right-of-way in Avoca. A siding 450 feet in length trailed from the southbound main track. This was the site of the former North Avoca station, but the name was now changed to simply "Avoca." A freight platform, 56 feet by 16 feet, was constructed at Hancock. Since the team tracks in Pittston were inadequate, the company developed a small yard east of its tracks near the Pittston station. Equipped with overhead trolley wire, the siding had a capacity of nine cars and could be expanded if necessary. Another siding of 270 feet was built near the powerhouse in Scranton for a business which planned to locate there. By 1909, five acres were leased to the Spruks Brothers Lumber Company, which would become an important shipper. The Scranton Ice Company also leased a small piece of land at the terminal, and a few additional industries located at sites in Wilkes-Barre.

Conn, in his efforts to emphasize freight, refused to join the Pennsylvania Street Railway Association, claiming that as the L&WV underlying companies were chartered under the general railroad laws of Pennsylvania and had no trackage in public streets, the carrier was really the equivalent of a steam railroad. L&WV officials frequently made the same assertion in their efforts to obtain interchange agreements with other railroads.

By 1911, gross annual revenues from freight amounted to $60,000. Some of this was hauled in the express cars, which could pull up to three standard freight cars, and the remainder was carried in regular freight trains operated by a four-man crew, with one of the two electric locomotives providing the power.

In summary, however, industrial development was disappointing. The number of companies which actually built facilities was considerably fewer than Conn's rosy prediction in 1906. Most were minor shippers who filled an average of less than one car per week.

Express and Mail. Effective March 1, 1905, the L&WV implemented an agreement with the Adams Express Company for the carrying of packages and other shipments. Adams Express was permitted to use stations and cars of the L&WV which were in regular freight service. Also, the Laurel Line would run a special car, leaving Scranton at 6:10 A.M. and, returning, leaving Wilkes-Barre at 9:35 A.M. In an emergency, Adams might load important packages on regular baggage cars. The Pennsylvania Railroad was reluctant to have any dealings with the electric line, and for some time it refused to permit its express cars from New York and Philadelphia to be sent over the Laurel Line. After four months, Conn expressed displeasure at the amount of business which Adams Express was providing. He unsuccessfully argued that it should increase its payments to the Laurel Line.

Relations between the Adams Express Company and the L&WV remained strained. One reason was Conn's refusal to permit Adams to load special shipments on the company's regular baggage cars. For example, in October 1908, the Adams agent in Wilkes-Barre complained to his superintendent in Philadelphia that he received a shipment of meat consigned to Armour and Company in Scranton, which he attempted to place on the 2:30 P.M. train. The L&WV agent refused to permit this and insisted the 20 boxes of sausage wait until the 5:40 P.M. freight. This placed it in Scranton too late for delivery that day. Conn's position was that regular combine or baggage cars could be used by Adams only if they sent a special messenger to accompany these shipments.

He also refused an Adams request that the express car hauled from Scranton be placed in the Laurel Line station at Wilkes-Barre instead of in the Lehigh Valley station. That line charged Adams $10.00 per day to set out a car, but Conn maintained that the L&WV did not handle express at its passenger station. So the matter was dropped. Even with these disagreements the contract remained in force during this period. Gross earnings from express were modest, ranging from $4,100 in 1906 to $5,200 in 1909.

The Laurel Line tried several times to secure a contract with the post office department for the carrying of mail. Representatives of the Railway Post Office service inspected the property, but because the electric cars lacked suitable facilities, the L&WV was judged to be unqualified and no contract was awarded.

Right-of-Way. As with all railroads, maintaining the stations and the right-of-way was an endless task. Most of the land used by the line was originally acquired under condemnation proceedings or the threat of such action. The right of a company chartered under the general railroad laws of the Commonwealth to use electric power rather than steam locomotion was challenged to court. Eventually, the Supreme Court of Pennsylvania upheld the right of the underlying companies to utilize electricity under their "steam railroad" charters. These state laws also prohibited a "general railroad" from physically connecting with street railways or using street trackage. For this reason, the L&WV was constructed on private right-of-way with large terminal facilities.

In its completed form, the L&WV had a total of 22.63 miles of first track, 20.74 miles of second track and 6.32 miles of yard track and sidings for a total of 49.69 track miles. The company claimed that it had built a first-class railroad, but some miscalculations and errors soon became apparent. For example, in 1907 it was necessary to replace all the telephone wires with a heavier type because the original wires were too light and broke under the weight of snow. The company forgot to install telltales at low points and only belatedly took steps to make sure its employees were familiar with these danger spots. A more serious omission was the total absence of a signal system on the line except for the staff machines at the tunnel, and a few manually operated semaphores. Eventually, the L&WV would pay a heavy price for this neglect.

Modifications and improvements were made to stations during this period. When the new MU cars went into service in 1904, platforms were extended to a length of 180 feet to accommodate three-car trains. After the Lackawanna Railroad built its new station in Scranton, Conn proposed that both companies share the expense of constructing a subway under their tracks, which would provide a direct connection between both terminals. Vice President B.D. Caldwell of the DL&W rejected this plan, saying the Laurel Line would get all the benefit and that passengers seeking DL&W trains would find their way to its station by the longer street route in any event. Conn asked Smith to see if he could persaude the DL&W to go along. He, too, failed and the project was canceled. Nonetheless, relations between the two companies were generally amicable. When the DL&W wanted to buy a strip of land 678 feet long adjoining its right-of-way in Scranton, the L&WV made the sale for $750.

Despite its perennial need for additional revenue, the L&WV rejected many proposals for installing vending machines in its stations. Conn felt they were not necessary for the "comfort of our patrons," and they took up valuable space. The station buildings were reasonably maintained and in 1907 the property was generally spruced up. Buildings were painted and crossings and bridges were re-planked. The L&WV was responsible for maintaining the bridges at Moosic and Pittston used by the Erie Railroad and for the structure at Wilkes-Barre used by the Wilkes-Barre and Eastern Railroad. These and all steel bridges were repainted in 1911 at a cost of more than $3,000 and local stations and fences were repainted at a cost of $1,200. The big trestle at Avoca alone required 382 gallons of paint.

On May 11, 1911, the name of the Hancock station was officially changed to Plains at the request of the solicitor of Plains Township. A request to move the Pittston station from Market Street to Broad Street was ignored. A new station was opened at Virginia, about one mile south of Connell Junction, in 1906. The company hoped that industrial development in the isolated area would provide passengers, but few materialized. Little was lost, however, since the minimal investment resulted in two unsheltered platforms 61 feet long, separated by a fence.

Maintaining fences along the right-of-way was a constant expense, but, because of the third rail hazard, they could not be ignored. Much of the fence was barbed wire which had to be continuously checked and repaired. Four miles of this fencing were installed in 1909, and in 1911, 25 miles of barbed wire were purchased. Rail was replaced as needed, especially on curves and other points of high wear. Conn tried to cut costs by buying used or light rail for sidings. He purchased 3,000 feet of 60-pound rail in 1908 for this purpose, only to discover, to his chagrin, that it was unsuitable for the power return and had to be replaced with 75-pound rail. More than 1,500 feet of 90-pound steel was laid as guard rail on all curves near the Susquehanna River between Prospect viaduct and Hollenback crossover near Wilkes-Barre as protection against derailments.

In 1909, more than 5,000 feet of new track was laid in sidings and 4,400 feet of rail installed on mainline curves. And more than 6,700 feet of trolley wire was installed; this was required partly because of carelessness by the trainmen leaving the third rail sector at Wilkes-Barre. Conductors, especially at night, failed to put the trolley pole carefully on the overhead wire, thus causing damage. For some strange reason, a seven-foot-high wooden fence was erected along the canal tracks in Wilkes-Barre in 1905 and although it was painted a soft color, many considered it an eyesore. When it was replaced five years later by a screen fence, reinforced with a hedge, the company earned the praises of the newspapers and residents of the city.

The Coal Menace. An ongoing problem, which ultimately cost the company considerable sums of money, was the presence of coal under the right-of-way. Approximately 14 of the

The Scranton skyline was drastically altered in 1907 and 1908 by the construction of the new passenger station for the Delaware, Lackawanna & Western Railroad, located adjacent to the Laurel Line station. Today, the structure is the ***Hilton at Lackawanna Station*** and also the focal point for Steamtown excursion trains to Moscow. The slumber of the guests is not disturbed, for through freights vanished from the ***Road of Anthracite*** in 1980.

George Arents Research Library, Syracuse University

20 miles of main line were underlaid with from one to 12 veins, varying in thickness from three to 14 feet each and in depth below the surface of from 15 to 1,100 feet. Legally, the coal was owned by various coal companies, who had the right to remove it. Therefore, the Laurel Line either had to purchase the coal, which was very expensive, or make temporary arrangements to run its trains during the mining operations.

The removal of anthracite consisted of two steps: "first mining," which removed about two-thirds of the coal and left one-third in pillars for surface support, and "robbing" or the removal of the pillars. The first step caused no problems, but the "robbing" was inevitably followed by the fall of the roof of the tunnel, which then caused a surface collapse.

The company dealt with surface disturbances in several ways. In some cases, temporary tracks had to be laid around the area being "robbed." For example, in 1909 the Hillside Coal and Iron Company informed the L&WV that it was about to begin robbing pillars in the vein under the tracks between Dupont and Williams Street in Pittston, a distance of 2,500 feet. Since the roof of the tunnel was only from eight to 20 feet thick and consisted of gravel, the surface would fall.

Conn rejected the solution of stopping trains on either side of the track and having passengers walk over this section because it would ruin business. Instead, he ordered a temporary track of 3,000 feet laid 100 feet east of the main line until the work was completed. To prepare the roadbed, a DL&W steam locomotive was rented for a few days. In other instances, the "robbing" was confined to a small area. Watchmen were posted and slow orders issued. As the surface fell, it was immediately filled in with the result that trains experienced only minor delays. Occasionally, the L&WV purchased the pillars and flushed the tunnel with silt through bore holes to prevent disintegration or theft of the coal.

Whatever the method, the Laurel Line expended considerable amounts of money. In 1910 alone, the L&WV paid the D&H Coal Company more than $1,400 for 800 feet of pillars and spent nearly $12,000 to move its tracks at Butler, which along with other expenditures elsewhere, totaled more than $20,000. In the following year, nearly $11,000 was spent for mining adjustments. These totals were charged to operating expenses and they obviously contributed to the company's financial problems.

Another persistent issue was safety at the six grade crossings in the city of Wilkes-Barre. Two of these streets carried the cars of the Wilkes-Barre Railway Company. An agreement made in 1910 provided that each company split the costs of maintaining the crossing frogs although the Laurel Line crews would do the actual work. Growing automobile traffic posed more of a hazard. The company employed watchmen at all of the intersections to flag traffic as cars approached. Usually they performed well, but sometimes irate motorists complained about their failings. One P.A. Meixell wrote to Conn in 1907 accusing the watchmen of carelessness. He said, "These fellows seem to be lazy and seek the shade rather than to be alert to their duties, and do very little signaling at all. They need stirring up. . . ."

Conn not only apologized but asked Mr. Meixell to act as an informer and report any further derelictions of duty. Soon, the Wilkes-Barre city council became involved and passed a resolution in August requiring the company to install gates on the grounds that the flagmen could not be seen adequately from the crowded street. Conn was anything but enthusiastic about the prospect of still another financial expenditure and since the gates would have to be manually operated by the watchmen, no savings would be effected. After consulting with an attorney, who felt the resolution was illegal, Conn simply dragged his feet and did nothing.

As there were many railroads entering the city, involving dozens of grade crossings, the question was expanded to cover all of these lines. The board of engineers, representing the railroads and the city, undertook an extensive study to try to eliminate grade-level crossings. The first plan proposed for the Laurel Line would have required the construction of an elevated structure to carry the tracks from the river to a proposed union station. The cost of this was prohibitive and a second option was considered. This would permit the L&WV to retain its terminal but called for an entirely new entrance, which would cost about $450,000. It was assumed the city would pay part of this and that a certain salvage value could be secured from the canal right-of-way. An additional advantage would be the creation of a terminal company with tracks used jointly by all the railroads. As the L&WV had at that time a connection only with the Lehigh Valley Railroad, the proposal would aid in the growth of freight traffic. However, after much study and talk, nothing materialized.

Personnel. Pay scales and rules of service for the L&WV were on a par with those of similar electric roads elsewhere. Trainmen wore standard blue uniforms, with white collars, dark ties and black shoes. When hired, they were placed on the extra list until a permanent vacancy occurred, when it would be filled on the basis of seniority. Periodically, all runs were "listed up" and reassigned to trainmen in the order of their seniority, except if the superintendent, for sufficient reason, ordered otherwise. If layoffs were necessary, they would be made from among "the least competent men among the more recently employed." Provisions were made for hearings and appeals in the case of dismissals. Passenger trains carried a motorman and conductor, while freight trains of two or more cars required a motorman, conductor and brakeman.

Conductors and motormen received 22¢ per hour for their first year, 23¢ per hour for their second year; and 24¢ per hour thereafter. In contrast, Vice President Conn was paid $500 per month and Superintendent Wilson and Traffic Manager B.F. Wiley received $300 each per month. By 1909, motormen and conductors were earning 34¢ per hour, while brakemen were paid only 22½¢ per hour. Firemen in the power plant got 25¼¢ per hour. Five years later, in 1914, wages were slightly lower. Conductors and motormen were pegged at 33½¢ per hour and brakemen, 26¼¢ per hour.

Drinking, on or off the job, was frowned upon by management. Superintendent Wilson found it necessary to send a circular to all employees in 1905, stating that there had been several cases where employees ". . . under the influence of intoxicants . . ." had ". . . aggravated the offense by appearing on the company's premises . . ." and even reporting for work while in such condition. In each instance, the man claimed he had taken the liquor "as medicine" and consequently was not discharged. Wilson warned that ". . . intoxicating liquors and drugs have no place about a railroad . . ." and even use at home was not favorably regarded. If it were necessary to take as a medicine, ". . . home is the place to use it, and home is the place he must stay until the effects have passed."

This stern new policy was soon tested. In September, one Frank Rafferty, a trainman, reported for duty under the influence of liquor and was promptly suspended, despite his excuse that he was taking it as a medicine. When his suspension expired, he refused to report for duty on the grounds that he wanted to spend time in Hazelton. As a result, he was fired. He appealed to the union and the second vice grand master of the Brotherhood of Railway Trainmen interceded with President Smith in his behalf. Smith pointed out that drunkenness could not be tolerated and the best he could do would be to place Rafferty on the extra list. Conn was more severe than Smith. Two conductors, who were caught not registering fares, were fired by the vice president. However, when a committee of employees called on the president, he overruled his subordinate and ordered them reinstated because it was a first offense.

Accidents. Conn was probably justified in his concern about his employee's actions, since minor accidents occurred fairly often. For example, in February 1905, the 3:30 P.M. train from Scranton experienced trouble with its circuit-breaker at Rocky Glen. While the motorman was trying to make repairs and the conductor was walking back to flag, the following train rounded the curve and struck the stalled car. Fortunately, it was going only about five miles per hour when it hit and damage was slight. More serious was a derailment in July when a Dunmore car left the track south of Nay Aug Park and partially turned on its side. Four of the 20 passengers were injured slightly. The accident was blamed on the motorman for rounding the curve too fast.

This view taken in 1904 looks north at the "permanent" Rocky Glen station. Note the young boy, like most lads of his age, is gingerly placing his foot on the outer rail. The station was the only access to the park at this time.
Edward S. Miller Collection

In the first two months of 1905, a series of mishaps struck the line. Within a space of three days, there were four derailments. The first two occurred at the spring frog of the switch leading to the Pittston freight yard when the motor trucks of northbound passenger cars jumped the track, damaging the switch. The latter two took place at switches in the Scranton Terminal yards in a similar manner. In the following weeks, two more cars derailed at the Pittston crossover. Superintendent Wilson blamed excessive wear on the wheels of the motor truck, since the treads were worn somewhat hollow and the flanges were uneven.

A potentially serious accident was averted by quick action on the part of crewmen in 1909. Car 103 was descending the hill at South Scranton when an object flew up and struck the bleed cock on the auxiliary reservoir, causing the air to leak. By manipulating the hand brake and using the motors, the motorman kept the car under control until it stopped on the level ground. The crew then plugged the hole in the reservoir with a piece of wood and managed to reach Wilkes-Barre safely.

Not so lucky was train number 9, car 17, which ran away down the Scranton hill on June 19, 1911. The train was due in Scranton terminal at 7:11 A.M., but as it proceeded down the hill after traversing the Laurel Junction loop, the motorman lost control as the brakes failed. He followed the emergency procedures outlined in the employee timetable of opening the circuit breaker cutout switch and reversing the controller, thus creating a dynamic brake. Although this slowed the car somewhat, its speed of about 50 miles an hour and the almost four percent grade were too much. When the car swung into the terminal loop, it left the rails and turned on its side onto the station platform. Fortunately there were no serious injuries. Later, a derail was installed on the hill to prevent runaway trains from reaching the loop.

Unfortunate citizens sometimes carelessly crossed the tracks in the face of oncoming trains. For example, a former justice of the peace of Pittston was killed in October 1905 near the Hancock station. A number of people claimed injuries when alighting from the front ends of cars. This prompted Superintendent Wilson to issue an order that passengers not be allowed on the front platform while the train was moving. One motorman suggested that stretchers and canvas be kept at Scranton, Pittston and Wilkes-Barre to cover persons killed by cars. Conn, aware of the negative reaction which the public would have, ignored this suggestion.

Westinghouse Recognizes Reality. Whatever his original motives may have been in sponsoring the construction of the Lackawanna and Wyoming Valley, by 1907 George Westinghouse began to have doubts about the financial return on his investment. Early in the year he asked George Smith to furnish him with a complete statement on the origin, construction and operation of the line. As usual, the management of the road depicted its affairs in an optimistic, but misleading, light. Smith gave a brief background and painted a glowing picture of the company at that time.

In referring to the properties at Scranton, Pittston and Wilkes-Barre, Smith repeated several times that they could not be duplicated at any price and were worth many millions of dollars. He admitted the growth of freight traffic had been disappointing and blamed it on misunderstandings and criticism by certain steam railways. After reporting that freight earned a gross revenue of $2,000 per mile per annum and passenger traffic earned $20,000 per mile per annum, he predicted the development of local industries and expanded interchange agreements would increase earnings dramatically ". . . in the near future."

He suggested that ". . . a closer alliance with one or more of the important through lines in this territory would establish the earnings of this property in a much shorter period." He asserted the property's actual and prospective value to lines such as the Pennsylvania, Erie, Lehigh Valley, Delaware and Hudson and New York, Ontario and Western ". . . cannot be even approximately measured by the cost of the present lands and improvements."

Despite the optimism of Smith's report, which was based on information and attitudes presented by Conn, Westinghouse executives reached a more realistic conclusion. Although the line had an impressive operating ratio and net earnings, it could not cover its initial capital investment. As a result of this analysis, George Westinghouse seized an opportunity presented by a news article in the *New York Tribune* on July 1, 1907, to take the first step toward relinquishing his interest in the Laurel Line. The article reported that both the Central Railroad of New Jersey and the Delaware and Hudson were considering electrifying their lines in the Wyoming and Lackawanna valleys with a third rail system.

Westinghouse had an article drafted to be anonymously placed in the New York newspapers extolling the value of the Laurel Line, especially to the Delaware and Hudson. He forwarded this to Smith, who promptly sent it to Conn with a confidential letter asking him to place it with a responsible correspondent in Scranton for publication as a news item. He indicated that Westinghouse wanted ". . . this dispatch to appear as written and hopes you will be able to so arrange without, of course, our interests being known in the matter." Two days later, Conn reported he had placed the material, ". . . with a party who assured me that he could have same appear in two or more of the New York dailies and in a way that its origin would be unknown."

The intentionally misleading article was datelined Scranton and described the L&WV as "easily the most popular among all the transportation lines in this territory. . . ." Referring to the proposed electrification plan of the CNJ and D&H, the dispatch claimed that they were false reports attempting to injure the prestige of the L&WV in financial circles. The article went on to claim that a merger with the D&H was probable and "logical." The L&WV had valuable terminal facilities while the D&H did not, and if the steam road were to acquire the electric line it could use its own tracks exclusively for freight and let the Laurel Line carry all the passenger business. In addition, the D&H could electrify its Carbondale division, using power generated by the L&WV, and would gain new traffic from the industries along the Laurel Line.

Finally, the article noted, the Delaware and Hudson would secure a strategic advantage by ". . . the final exclusion of the Pennsylvania from Scranton as well as the exclusion of the Ontario and Western from Wilkes-Barre and a possible connection with the Pennsylvania at that place." The reaction of the D&H managers is unknown, but they were evidently unimpressed with the prospects offered by the control of the Laurel Line.

Undeterred by the failure of this first effort to entice a steam railroad to acquire the electric road, Conn tried again late in 1908. He took advantage of a meeting with Kerr of the New York, Ontario and Western to bring up the matter of the attractive prospects of the Laurel Line. Conn told Kerr that surveys had been made concerning the possible extension of the L&WV from Dunmore to Throop and a connection with the O&W. He pointed out that some of the hundreds of cars of freight transferred by the PRR to the New Haven at Jersey City could then be routed through Wilkes-Barre over the Laurel Line to the O&W and, eventually, the New Haven. If either of the giant trunk lines (the PRR or the NH) objected to losing even the revenue for the 25-mile L&WV haul, they could easily acquire the electric road.

Kerr seemed "much interested," although he added he could not see any value in the O&W acquiring the L&WV, as it would merely put them in Wilkes-Barre, "which would be little gain." Conn then dangled the lure of additional coal traffic, pointing out that if the O&W acquired the Wilkes-Barre and Hazelton Railway, it would gain an entrance to the southern anthracite field. He told Smith he had not intended to bring up this matter but merely took advantage of a good opportunity. He felt it was ". . . perhaps the best solution of this situation," as the investment which the O&W would have to make to acquire the L&WV ". . . would be insignificant in comparison with the benefits which accrue to it. . . ." He concluded that if the O&W, backed by the New Haven, made the investment ". . . I believe that with the adjustment of our securities which their guarantee would make possible, the property would be in a position to show a good profit from operations." A good try, but nothing came from this scheme either.

Finally, an Acquisition Proposal. In contrast to these nebulous schemes of trying to interest steam railroads in the property, a concrete proposal came from the American Railways Company in March 1910. Newton Jackson, an official of the Philadelphia-based holding company, approached George Smith with an offer to secure control of the L&WV. He proposed that the American Railways refinance the company by issuing new bonds and stock. He also expressed an interest in another Westinghouse property, the Grand Rapids, Grand Haven and Muskegon Railway Company.

Smith did some checking on the American Railways Company (ARC) and discovered it had assets of about $13 million. He told Robert Mather, chairman of the board of the Westinghouse Electric and Manufacturing Company, that the

With windows open to catch the summer breezes, a northbound train rumbles over the Spring Brook Railroad and the stream of the same name. At the left end of the bridge is the Moosic station, after its relocation from the original Powder Mill Road site. In spite of having almost every seat filled, even in 1907 the system was unable to earn enough to carry its debt load.

Dr. Edward Steers Collection

purchase of the L&WV ". . . would appear to be a rather large transaction for this company to handle . . ." but he was making confidential inquiries in Philadelphia concerning the American Railways Company. H.B. Gill, who was both general counsel of that company and a director of the L&WV, indicated to Smith that the firm was conservatively and soundly managed. Its principal officers were men of wealth and reputation in the city of Philadelphia who had interests in transit and other companies. The American Railways Company controlled the street railway systems of Scranton, Altoona and Johnstown, Pennsylvania, Lynchburg and Roanoke, Virginia, and several other properties. It paid six percent dividends on its stock on a regular basis.

Smith and Mather met with Jeremiah J. Sullivan, the president, and other officials of the American Railways Company. A tentative agreement was reached whereby Westinghouse would transfer all the outstanding bonds and most of the stock of the L&WV Rapid Transit Company to the ARC in exchange for a new issue of $5 million first mortgage, 4½ percent, 50-year bonds guaranteed by ARC and $2,500,000 in 50-year income bonds. An additional $2,500,000 in income bonds and any new stock issue would remain with the ARC. The deal was subject to approval by the ARC after it examined the physical facilities and financial records of the L&WV. In substance, then, Westinghouse was willing to accept $7,500,000 for the L&WV.

Smith authorized a representative of the ARC to examine the properties of his company, but he asked that the matter be kept secret, which Sullivan was willing to do. The representative who examined the physical facilities of the L&WV was unimpressed. He reported a considerable amount of money would have to be spent within the next two years on the roadbed, rolling stock and powerhouse. Consequently, Sullivan informed Smith on April 7, 1910, that the physical condition of the road was such that it was unnecessary to examine its financial status. Therefore, he declined to purchase the property.

For a brief period negotiations continued with both the American Railways Company and the L&WV pushing for more favorable terms. Neither party was willing to compromise and the offer to purchase was officially withdrawn on June 16, 1910. It was apparent to both Westinghouse and the ARC that the Laurel Line was a losing proposition. Approximately $9 million had been expended to construct a line less than 23 miles long. Total annual interest charges on the outstanding mortgage bonds were more than $384,000, while net earnings in 1908, 1909 and 1910 averaged about $248,000 per year. In addition, maintenance costs were likely to increase, while earnings would remain steady at best. With the failure of these negotiations, the Westinghouse companies would have to seek other ways to salvage their investment.

Financial Reorganization Looms. While from an operating point of view the L&WV might be considered a success, financially the line was a failure. There were growing indications of a shortage of operating funds. By 1909, the company's correspondence shows a consistent pattern of refusing to pay suppliers. It became standard practice for Conn to complain the bills were incorrect and to demand a reduction in the amount. Often Conn engaged in correspondence over a period of months with increasingly irate vendors over trifling amounts, while at the same time demanding immediate delivery.

For example, a chemist who analyzed solder submitted a bill for $25, which Conn claimed was exorbitant. In another instance, in an act of charity, Conn agreed to pay the burial expenses of a vagrant who was killed by a train at Moosic. The undertaker estimated it would cost $10, but Conn was outraged when he received a statement for $40, which he refused to pay. Eventually, a settlement for $30 was agreed upon. At the same time, the company maintained a stubborn position in handling accident claims and often appealed to higher courts when verdicts were considered unfavorable.

Unbelievably, during the period of Westinghouse control, the books of L&WV were not audited through 1911. Only when a financial reorganization was imminent did the executive committee demand an audit, which was completed by 1912. Thus, the Westinghouse management followed the dubious lead of George A. Lee during the construction period when he had similarly refused to reveal the financial state of the line.

Considering the magnitude of its investment in the L&WV, Westinghouse should have been very concerned about its financial condition. As of February 1, 1909, the Westinghouse companies held 6,349 bonds valued at $6,320,042 out of a total issue valued at $7,566,042. Of the 65,000 shares of stock worth an estimated $1,840,322, Westinghouse's Security Investment Company owned 62,036 shares valued at $28.23 per share, or a total of $1,751,502. Even worse, the L&WV was not earning a sufficient revenue to cover both its operating cost and interest on its bonded indebtedness. As of January 31, 1911, the cumulative deficit amounted to $1,267,337. Most of this was covered by the Westinghouse firms, which deferred presenting their coupons for payment in order to prevent default.

In 1907, a brief financial panic, called variously "The Bankers' Panic" by President Theodore Roosevelt or "Roosevelt's Panic" by the bankers, gripped the market. As a major investor in various enterprises, Westinghouse's companies were under growing pressure. By July of 1908 the Mercantile Trust Company, which held as collateral for a loan to Westinghouse the securities of the L&WV Rapid Transit Company and other properties, demanded more collateral because of the falling market values of the stock and bonds. Of the total collateral of nearly $9 million, $6 million was represented by L&WV Rapid Transit Company first lien consolidated mortgage bonds.

It was apparent that the L&WV was regarded as an overcapitalized corporation, whose solvency was being maintained by Westinghouse. In some respects, the almost total Westinghouse dominance of the L&WV securities was an advantage, since minimal pressure developed from the public at large over the growing deficits of the Laurel Line. When Howard Dunham of Philadelphia tried to sell his 50 shares of stock for only $10 a share, he could find no buyer. Smith explained there was no market for the stock, since most of it was owned by Westinghouse, which certainly did not want any more of it.

Although Westinghouse was not getting any return on his investment, at the very least he expected the L&WV to buy all of its material and supplies from his firms. For instance, in 1910 Conn responded to Smith's request for a list of purchases made from other than Westinghouse suppliers with a statement that only about $110 had been expended for items

A rare three-car train, heading south, is ignored by a pensive young man in a rowboat on the Rocky Glen Lake. The Stephenson-built car in the center of the train is distinguished from the two slightly older ACF-built cars by its lower hanging truss rods. By this date, 1908, the type L's had been rebuilt as double-ended, multiple-unit control cars. *Edgar J. Boland, Edward S. Miller Collection*

which were also available through Westinghouse operations. Thus, indirectly, Westinghouse did receive some benefits from its ownership of the L&WV, but hardly enough to cover its losses.

Certain internal changes in the management structure were made. At the end of April 1907, George A. Lee, one of the original promoters and prime mover of the Transit Contract Company, submitted his resignation as a director and member of the executive committee of the L&WV Rapid Transit Company. This was shortly after he and his affiliated companies released the L&WV from a claim for $200,000. As part of an agreement negotiated on April 1, Lee's Transit Finance Company sold to Westinghouse's Security Investment Company 10,000 shares of stock of the Rapid Transit Company; Security Investment, in turn, delivered to the Transit Finance Company $300,000 in L&WV bonds. These bonds were placed in trust with the Standard Trust Company of New York for a period of three years, after which Security Investment would either repurchase them for $300,000 or find a buyer. For reasons that are now obscure, a dispute developed in 1910 when Security Investment refused to buy back the bonds. The matter was taken to court, but in the meantime Security Investment continued to pay interest.

More changes were in the offing. At the end of August 1911, George Smith resigned as president and director of the L&WV Rapid Transit and subsidiary companies. Pending an election by the board, Calvert Townley, assistant to the president of the Westinghouse Electric and Manufacturing Company, took control of the corporations. Although the relationship between Smith and Conn remained cordial, the vice president lost no time in asking Smith to return his annual passes on various railroads. When Smith complied, Conn secured replacements in the name of Calvert Townley, who was elected president on September 22. Also replaced at the time were Treasurer Carl Vail and a few others.

As early as 1909, it was apparent to Westinghouse officials that a reorganization of the L&WV was necessary. When Smith learned the Security Investment Company was considering authorizing the Standard Trust Company to sell some of the stock of the Lackawanna and Wyoming Valley Rapid Transit Company, which was held as collateral for loans, he protested. In a letter to C.S. Gray, vice president of Security Investment Company, Smith pointed out that a tentative plan for reorganization without judicial sale had been prepared. If any other stock were sold, it would only have to be repurchased at a higher price. On the other hand, if there were a reorganization by foreclosure, the stock would probably be wiped out and therefore should not be sold at all.

Gray admitted the executive committee had contemplated selling some of the stock at a minimum price of $13 a share. He expressed surprise that anyone would be interested in purchasing the stock and he thought it very unlikely that the prospective buyer would complete the transaction. After personally consulting with Smith, the executive committee decided not to authorize the sale of stock.

By 1912, the situation had reached a critical stage. On August 1, the company was forced to default on its interest payments on its first lien and consolidated mortgage bonds. Although a partial payment was made in subsequent months, by early 1913 there was still a large amount of interest due, which the company was unable to pay. Therefore, under the terms of the mortgage, the Guarantee Trust Company (successor to the Standard Trust Company) declared the principal and interest of the first lien and consolidated mortgage bonds due and payable. Westinghouse, which was the prime owner of the consolidated bonds, had been deferring interest due it since 1908 and continued that tactic in order to arrange a friendly reorganization through the courts. Interest on the bonds held by the public was paid.

Therefore, the suit filed by Guarantee Trust was merely a device to reach a legal settlement acceptable to Westinghouse and the other stockholders. Essentially, the plan involved the redemption of all the bonds of the various underlying companies and their consolidation and merger under a new company to be called the *Lackawanna and Wyoming Valley Railroad Company*. This corporation would issue $5 million of 5 percent first and refunding bonds, of which $2,638,000 was held in the treasury as reserve against the retirement of the collateral trust bonds and for future expansion and $2,362,000 was issued at once. These bonds were secured by a mortgage which was a first lien upon all of its property and franchises. A total of $1,250,000 of cumulative preferred stock and $3 million of common stock were issued, making the total obligations of the new firm $6,612,000. The fixed charges, which involved interest payments on the bonds and dividends on the preferred stock, were $237,500. Present earnings exceeded the figure and it was expected the company would therefore earn a profit.

The new securities were distributed among the holders of the old bonds. Each holder received the new paper in the following proportions to the face value of his present holding: new bonds, 31.394%; preferred stock, 16.614%; common stock, 39.874%; total 87.882%. Since Westinghouse owned $6,074,000 of the original bond issue of more than $6,500,000 it was obvious that it sustained a substantial loss on its original investment. Westinghouse also paid all the expenses of the reorganization. Apparently the Westinghouse interests considered the loss worthwhile because they would finally be relieved of their responsibilities for operating and supporting the L&WV.

As a byproduct of the reorganization, Conn retired as vice president on July 31, 1913. Perhaps disillusioned with his experiences on the Laurel Line, Conn abandoned the railroad industry entirely and soon became president of the Giant Portland Cement Company of Philadelphia. He was succeeded by W.E. Higgins as acting vice president and general manager.

The United States District Court for the Southern District of New York sanctioned the reorganization plan in April 1913. The merger agreement was approved on May 26, 1913, and letters patent were issued on June 20, which canceled all prior leases. At the same time a new corporation was chartered under the laws of Virginia, called the Scranton and Wilkes-Barre Traction Corporation. This was essentially a holding company whose purpose was to acquire all of the bonds and stock of the new L&WV Railroad Company. It issued $1 million of 7 percent preferred stock and $2,398,000 of common stock. In addition, the traction company was authorized to issue $3 million of 5 percent refunding, sinking fund gold bonds.

For the time being, Westinghouse interests had control of the holding company, but they were seeking a buyer, which would end their involvement with the L&WV. In the following year, a syndicate headed by state Senator William C. Sproul of Chester and backed by the Dupont interests, the Wetherill family of Philadelphia and Bioren and Company, also of Philadelphia, purchased the Scranton and Wilkes-Barre Traction Corporation for $1,600,000, for which they received $1 million of preferred stock and $1,500,000 of common stock.

In retrospect, it appears that the Westinghouse interests made a poor investment in the L&WV. The basic problem can be traced to the extraordinarily high construction cost and the apparently slipshod accounting practices that were followed by the promoters. Another misconception was the belief that a high volume freight traffic could be developed which would utilize the line's double-track right-of-way and offset some of the fixed charges. Passenger traffic did increase and the line had become the prime mover of people in the valley region. It remained to be seen whether the new owners, favored with reduced interest payments, could make the Laurel Line a financial as well as an operating success.

TWELVE MONTHS ENDING JANUARY 31

	1905	1906	1907	1908	1909	1910	1911	1912
GROSS EARNINGS								
Freight	27,436	20,541	33,823	44,790	49,305	56,069	61,814	62,459
Passenger	334,204	378,513	410,923	450,206	483,007	492,623	513,639	519,851
Other	3,440	10,853	12,428	16,495	12,810	13,298	15,506	16,157
Total	365,080	409,907	457,174	511,491	545,122	561,990	590,959	598,467
EXPENSES								
Operation of Power Plant	41,942	40,993	32,077	34,785	53,013	54,714	50,110	56,445
Operations of Trains	85,677	98,437	105,480	118,370	125,969	127,254	130,732	129,016
Other	85,136	89,548	100,430	111,131	145,604	153,127	159,816	146,925
Total	212,755	228,978	237,987	264,286	324,586	335,095	340,658	332,386
NET EARNINGS (After Taxes)	151,300	169,088	203,757	241,047	212,116	217,881	240,680	256,167
OPERATING RATIO (% of Operating Expense to Gross Earnings)	58.27	55.85	52.05	51.60	59.54	59.62	57.84	55.54
TRAFFIC CARRIED								
Revenue Passengers	2,485,139	2,894,601	3,285,780	3,514,948	3,732,519	3,768,421	3,927,424	3,997,054
Total Tons of Freight	92,484	17,423	66,373	96,349	108,775	123,356	124,701	133,617
POWER								
Kilowatt Hr Power Used	4,903,484	5,846,066	6,855,110	7,806,663	7,784,486	8,168,681	8,276,852	
Cost per KW Hr (Mills)	8.55	7.10	4.68	4.45	6.81	6.69	6.05	

Seemingly in a remote gorge but actually within the city of Scranton, a short freight train clatters over Roaring Brook near the north portal of the tunnel.
Edward S. Miller

3

The World War I Era: Problems and Progress for the Laurel Line

IN THE SAME YEAR, 1913, that a new management assumed control of the L&WV, a new administration and party were taking power in Washington. On March 4, Woodrow Wilson took the inaugural oath, called a special session of Congress in which his Democratic Party held a majority and startled and pleased the nation by appearing in persòn before it to deliver his message, the first president to do so since John Adams. Wilson would develop a moderately progressive policy that encouraged a friendly regulation of business. But then domestic problems receded into the background after August 1914 when Europe was plunged into a titanic struggle known as the Great War.

Although the president was determined to maintain American neutrality and was narrowly reelected on that platform in 1916, the international climate was such that the United States was drawn steadily closer to the actual conflict. When the German government defied Wilson's strictures in the use of submarines, the president had no choice but to ask Congress for a declaration of war against Germany in April of 1917.

The conflict in Europe resulted in increasing demands on the American economy as agriculture, business, industry and transportation struggled to fill the growing orders for food and war materials from the belligerents. After the United States entered the war in 1917, the problems intensified as mobilizing and supporting an army, part of which was sent to Europe, increased the strain.

The war forced profound changes on American society. New taxes were imposed, especially on corporations and the wealthy. A federal War Industries Board tried to mobilize industry and establish priorities, and new distribution patterns, thereby often creating shortages of essential materials, especially iron and steel. The Fuel Administration encouraged the opening of new mines, which, in conjunction with other demands, placed an intolerable strain on the nation's railroads and steamship lines, which became incredibly snarled by the end of 1917.

The military draft and high-paying war industries created a labor shortage, while at the same time the new National War Labor Board and the War Labor Policies Board imposed unprecedented regulations on private industry, and by favoring shorter hours and higher pay scales increased business expenses considerably by 1918. Although the Laurel Line was only a minor cog in the nation's industrial machine, it would be hard pressed to meet the challenges imposed by the rapidly changing events of the second decade of its life.

A New Management Takes Control. The purchase of the L&WV by the Sproul syndicate and the establishment of the Scranton and Wilkes-Barre Traction Corporation to hold its shares and at the same time to offer its own securities on the market resulted in a much-needed managerial reorganization. There was a brief, transitional phase in 1914 when such Westinghouse men as W.E. Higgins retained operating control until the new personnel were ready to assume responsibility.

The first step taken by the Sproul syndicate, which was interested more in financial returns than in operations, was the hiring of a prominent firm of consulting engineers to supervise the activities, maintenance and direction of the railroad.

On April 24, 1914, a five-year contract was signed between the Lackawanna and Wyoming Valley Railroad and the firm of Ford, Bacon and Davis of New York City.

This organization was established in 1894 by Frank R. Ford and George W. Bacon. Pooling a meager $1,000, they formed an engineering firm and in the following year accepted George H. Davis as a partner. Their first major opportunity came in 1895, when they received a contract to electrify the Orleans Railroad in the Crescent City on the Mississippi. So successful was the firm in New Orleans that it soon became a leader in the field of building and managing electric lines.

Under the terms of the L&WV contract, the engineers were to ". . . supervise generally the regular operations of the company" and one of them, Ford, was soon elected chairman of the executive committee. In addition, the engineers were to act as consultants and to ". . . take charge of all work of construction, reconstruction, equipment, development and extension of the physical property of the company." Work amounting to more than $10,000 had to be referred to the

Two of the leading figures in L&WV affairs after 1914 were Patrick J. Murphy (LEFT, *from Lafayette College*) **and Frank R. Ford (RIGHT,** *from Ford, Bacon & Davis*).

engineers, while items costing less than that sum might be reviewed at the option of the company.

The engineers were also to supervise the purchase and installation of all materials involved in construction and to inspect the work to ensure it was properly done. For their services, the L&WV paid Ford, Bacon and Davis an initial fee of $5,000 per year, subject to increases on an annual basis. As consultants, Ford, Bacon and Davis would receive five percent of the cost of any construction work undertaken. All expenses and related costs were to be paid by the L&WV and any special work done by the engineers would be subject to an additional payment.

For all practical purposes, Ford replaced George Smith as chief executive of the company. Like Smith, he spent most of his time in his New York office and made only occasional visits to Scranton. Thus, the perennial problem of absentee management remained under the new regime. Since Ford could not afford to personally direct the line, he looked into the ranks of his employees for someone who could fulfill this function. He found there a man who was to guide and manage the affairs of the Laurel Line for the next 31 years, Patrick Joseph Murphy.

Born on March 13, 1870, at White Haven, Pennsylvania, about 15 miles south of Wilkes-Barre, to John and Abigail Murphy, the future president of the L&WV attended St. Patrick's School, White Haven, Harry Hillman Academy, Wilkes-Barre, and Lafayette College, Easton, where he graduated in 1898 with a degree in electrical engineering. Railroading was strong in the family; three of his brothers worked for the Lehigh Valley Railroad and a fourth for the Central Railroad of New Jersey.

Murphy, or P.J. as he was often known, secured a position with General Electric as a student in its Test and Construction Division. From 1901 to 1903, he was superintendent of power for the International Railways in Buffalo, after which he returned to General Electric for about a year. In 1905, he began working for Ford, Bacon and Davis, helping to construct power plants. The reasons for his selection as vice president and general manager of the L&WV in 1915 are obscure, but no doubt its proximity to his home appealed to him. Murphy was to spend the rest of his working life with the carrier, eventually becoming president and retiring in December 1946.

Of medium height, with bushy hair and eyebrows and with a somewhat portly build, Murphy always presented a neat appearance. A confirmed bachelor and a creature of habit, he spent weekends at the family's apple orchards in White Haven while maintaining his residence in the Hotel Sterling in Wilkes-Barre. Every weekday morning he boarded the same early train at the River Street station and read the *Wall Street Journal,* which the conductor had procured for him at the terminal newsstand. He usually ate his dinner at the Scranton station restaurant.

According to retired conductor Edgar Boland, Murphy was efficient, effective and always on the job. He was generally respected by the employees and he always had a slide rule at hand when discussing wage rates or other financial matters. He was an active member of such groups as the Kiwanis Club, the Elks, church societies and the Lafayette College alumni, often attending football games, dinners and social functions and serving as life president of his class. But his main interest was the Laurel Line, and its longevity, in contrast to that of many of its contemporary electric railways, may be partly attributed to his skill and ability.

Power House Problems. Although the property was only 10 years old, the machinery in the powerhouse required extensive modifications and improvements. A seemingly endless series of difficulties soon taxed the skills of the new consulting engineers. Some of the breakdowns may have been caused by unqualified workers employed by the company. For example, there was ongoing trouble with a new LaBlanc steam condenser built by Westinghouse. There was also concern about the danger of electrolysis damaging pipes, poles and cables of utility companies, particularly on the leg returning power to the substations.

The installation of new boiler water feed regulators, manufactured by the L.J. Wing Manufacturing Company, caused many disputes. They were delivered late and were not installed until sometime after they were received. Several engineers employed by the L&WV left the company and were replaced by new personnel who, when they finally installed the regulators, discovered they would not work and caused further damage. The L&WV had paid $300 to the Wing company but still owed a balance of $155, which it refused to pay until the trouble was corrected. A representative of the Wing company made several visits to Scranton and finally convinced the management that the fault was in the design of the plant rather than with the regulators. He recommended several costly changes, which would permit the apparatus to work. This the L&WV declined to do and also refused to pay the balance of the bill.

After extensive correspondence, the Wing company went to court. Eventually, the matter was settled privately and the

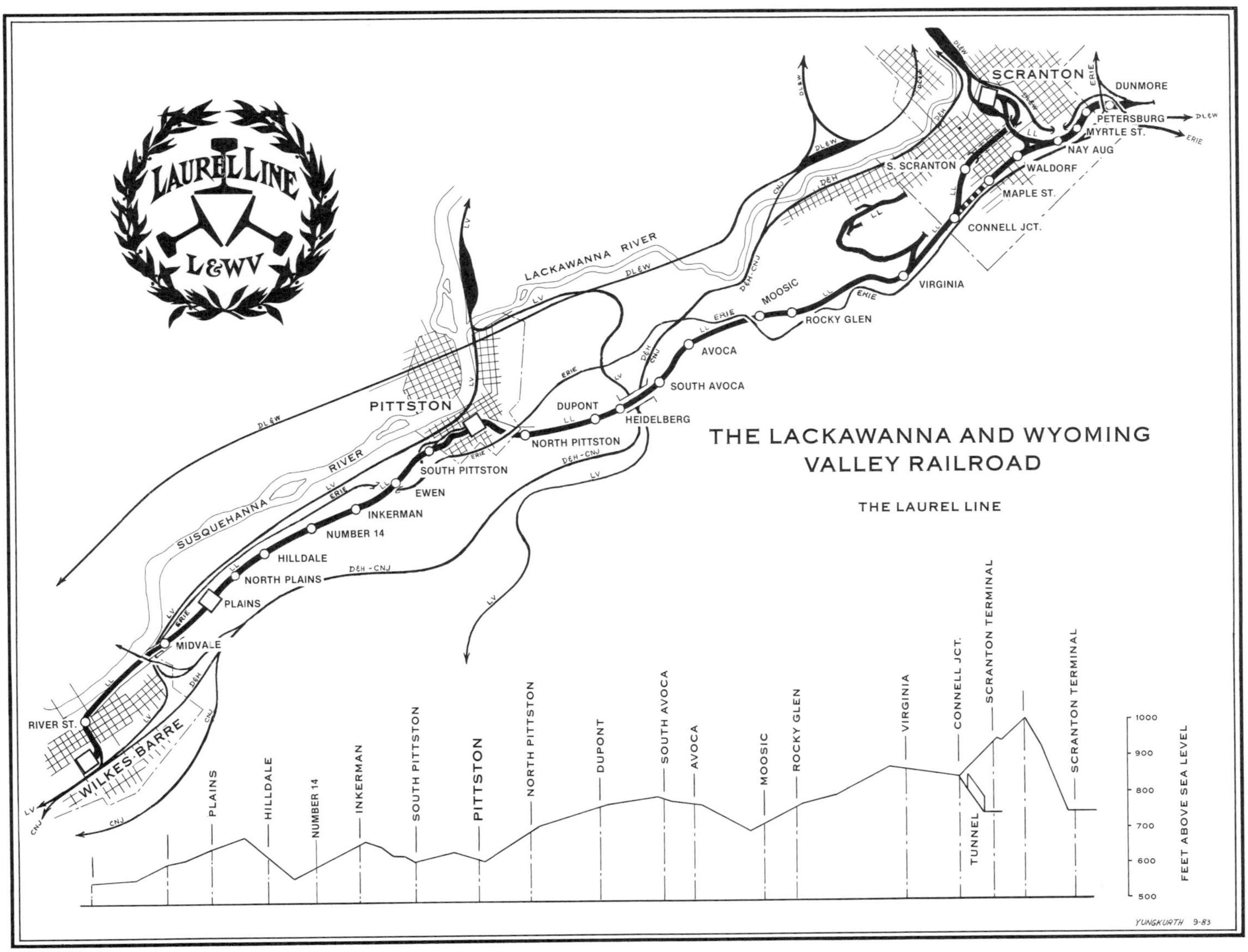

Two photographs of the Wilkes-Barre terminal facilities were published in a report on the L&WV issued in December, 1931, by Ford, Bacon & Davis. Scranton commercial photographer Horgan perched on the Wilkes-Barre & Hazleton Railway viaduct and aimed his camera both south (ABOVE) and north (BELOW). Unidentified items of railway interest are the Wilkes-Barre Railway Corporation's trolley viaduct over the railroads at the left edge of the upper photo, and the Lehigh Valley Railroad's Canal Branch between the LVRR mainline and the interchange track in the lower view.

Edward S. Miller Collection

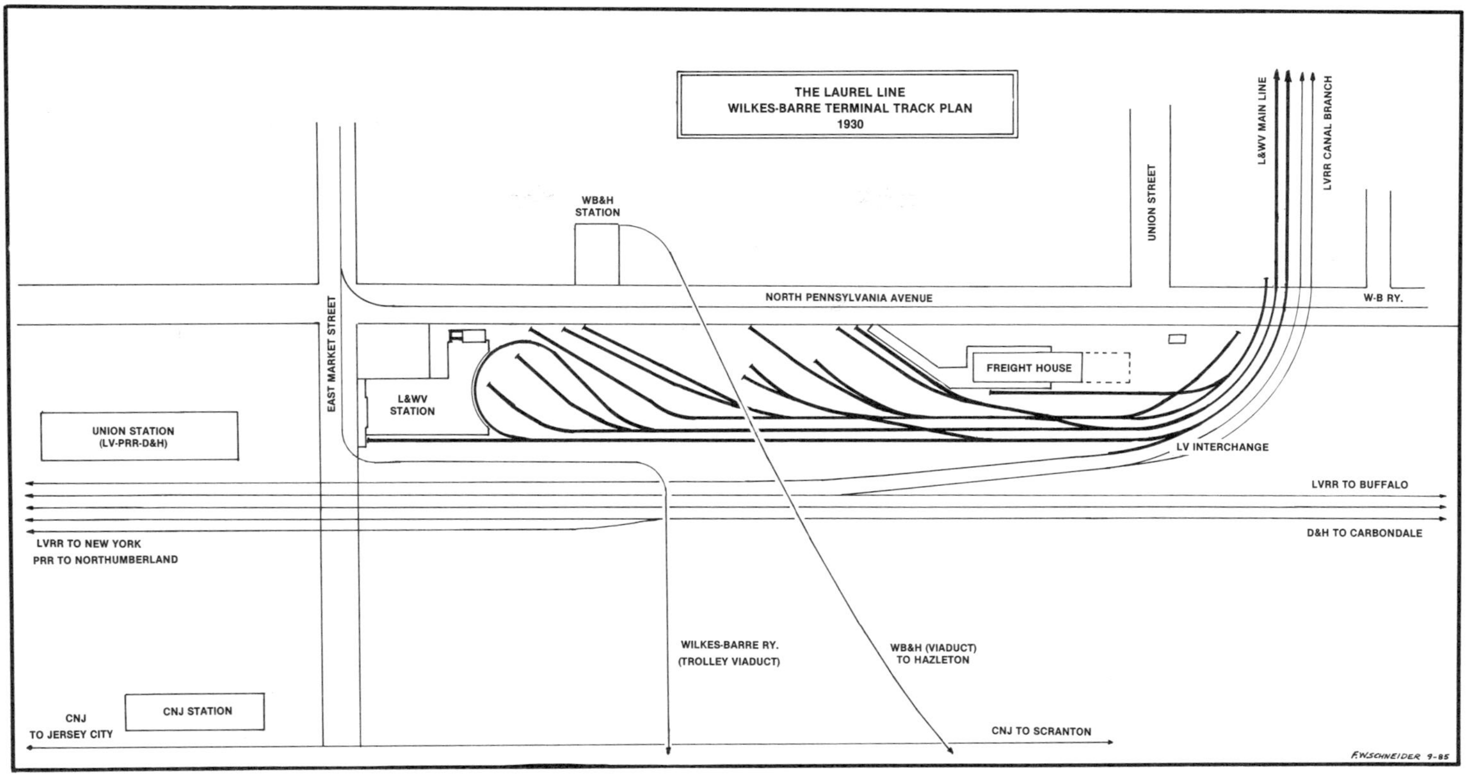

Laurel Line had to pay. Frank Ford concluded the debacle by rebuking Higgins for purchasing the regulators without a trial and without making projections as to the results. In the future, he reminded Higgins, definite specifications should always be developed. It was this lack of management expertise that had caused so many problems in the past and would continue to do so in the future.

Typical of the line's parsimonious attitude was a long-standing disagreement with a contractor who had supplied seven Reliance Safety Water columns for use on the Babcock and Wilcox boilers. They had not been used for two years, yet Patrick Murphy held the supplier responsible for what he judged to be defects in their construction and argued that the supplier ". . . should have made good these columns and not allowed the matter to drift on."

This was a minor issue compared to other difficulties that were plaguing the powerhouse. Major overhauls were necessary to keep the plant running but wartime restrictions required the use of priority certificates to place orders. These were obtained with the aid of the American Gas and Electric Company and the Scranton Electric Company in the fall of 1917 and included a 3,000-kw Westinghouse Electric turbine; a complete switchboard, together with three 1,250-kw, 60-cycle transformers and one 3,000-kw frequency changer set, all from General Electric; a 100-hp, 625-volt motor to drive the condenser circulating water cooling pump, also from General Electric; and a 60-hp turbine to drive a new 30,000-cubic-foot, forced-draft fan.

In addition, repairs were made to an older turbine. The railroad paid 50 percent of the cost when the equipment was received and the remainder over a period of 90 days. The frequency changer set and transformer equipment alone cost more than $55,000, while the turbine was priced at $46,000. In 1919, an additional $35,000 had to be expended for various repair parts, including a feed water pump, a 36-kw engine-driven exciter, superheaters for five boilers and a new coal conveyor system.

Operating the powerhouse proved to be so troublesome and expensive that the company developed ambivalent feelings about it. Several possibilities were weighed, including selling the entire operation to the Scranton Electric Company or making a joint arrangement with that concern for the sale of surplus power. The L&WV was already providing a limited amount of excess power to on-line industries, primarily coal companies, such as the Pennsylvania Coal Company's Number Three Shaft in Pittston and the Ewen Colliery in Jenkins Township. But this business, like the freight traffic, was disappointing and never developed to the level anticipated by the original promoters. One reason for this was the lack of a sufficient number of industries located along the line.

An effort was made to reach an agreement with the Scranton Electric Company. Under this plan, the L&WV would lease its plant to the American Gas Company, the parent firm, for $40,000 a year plus 10 percent of the gross revenue from sales to outside customers. In turn, Scranton Electric would supply all of the power needed by the L&WV. However, the utility was unimpressed with the Laurel Line's facilities and the high operating cost they entailed. Murphy explained that these costs were misleading because of recent repair expenses. The electric company would offer only $25,000 for the plant, which was rejected by L&WV management.

Still the idea would not die. The heavy investment carried by the railroad in its power facilities demanded some type of relief. When the L&WV tried to expand its commercial business to communities along the line, the Scranton Electric Company objected on grounds that the power company charter provided for sales only in the immediate vicinity of Scranton. Both courts and the Public Service Commission upheld the view of the electric company, which tolerated the sales to the collieries and a few industries because they were relatively minor. L&WV management concluded the difficulties and costs in obtaining a new charter were too great to make the matter worthwhile. Therefore, another attempt was made to reach an agreement with the electric company and this time success was achieved.

The L&WV was in a strong position for the moment, in the spring of 1917, because in 1915 it had acquired two large culm banks from the Pennsylvania Coal Company at a very low cost. It therefore could generate electricity for a smaller cost per kwh than the utility. Its right-of-way and power poles provided a convenient means for Scranton Electric to penetrate the territory south of the city. On the other hand, the L&WV would have to invest more than $100,000 in a new turbine in order to guarantee it could meet its commitment to supply power to the electric company. After many disputes, a joint agreement was reached whereby the L&WV supplied power during the day and early evening to Scranton Electric and received power in turn from that firm during the late night hours when its demand was low.

While these improvements and operating changes were underway, the company experienced ongoing difficulties with its generating of electric power. A series of complaints was voiced by William P. Jennings, president of the Pennsylvania Coal Company, which operated the Ewen Colliery, the Number Three Shaft, and other properties. On July 23, he reported that a two-hour power failure to a ventilating fan ". . . caused a serious condition which could have led to gas explosions and serious loss of life." He warned that if continuous service could not be provided he would have to ". . . make other arrangements." Although Murphy apologized and reassured Jennings, a similar failure occurred at another location in December. Murphy said the shutdowns were caused by inspections, but he promised to complete a second line to the Ewen Breaker, which would alleviate the problem. Before this could be done, another failure occurred in January and the exasperated Jennings simply noted, "This is for your information."

Other companies using L&WV power experienced service interruptions. The Delaware and Hudson Coal Company reported several breakdowns in February and March 1918. Even Senator Sproul, who was nominally president of the L&WV, became involved when he wrote to Murphy about the same time, saying,

> Pretty nearly everybody whom I see from Scranton and Wilkes-Barre has a grudge at me regarding the fact that our trains are running late. I have explained to them that it is due to powerhouse trouble and that we have been trying to do everything that we possibly could to correct this. . . .

Murphy was quite defensive in his response, indicating the community had been very fortunate in obtaining whatever service the line was able to give.

He blamed the difficulties on four sleet storms and the failure of 18 armatures on the car motors, as well as breakdowns in the powerhouse. He staunchly argued that the Scranton and Wilkes-Barre railways were ". . . down five or six times for each interruption we have had. . . ." For good measure, he added, "We have a very critical community about here and some people like to criticize in order to hear themselves talk." He concluded by telling Senator Sproul to advise his friends that conditions were far worse in other localities and only the fact that the L&WV had its own fuel supply prevented drastic cuts in service.

Despite Murphy's rather breezy dismissal of public complaints, there had been periodic power failures which caused hardship and inconvenience. In February 1918, power failed at the D&H Coal Company from 6:15 P.M. to 5:45 A.M. on the following day, stranding two men in the cage in the shaft, who nearly froze to death. Only two months later, a similar outage hit one of the shafts of the Pennsylvania Coal Company, although in this case power was restored in about an hour. Murphy apologized and said the company was doing everything it could to prevent recurrences. When a tie-line was installed between the Scranton Electric plant and the L&WV powerhouse, conditions improved and the number of failures declined. Still, the L&WV lost considerable goodwill among local businessmen because of the erratic nature of its electric service.

On the other hand, Murphy was quick to demand reimbursement when incidents caused by others inconvenienced the Laurel Line. In July 1918, an unfortunate boy fell into the pond behind the dam in Roaring Brook and drowned. To recover his body, the Scranton police opened the dam between 1:00 and 7:00 P.M., thereby shutting down the powerhouse and the railroad. Murphy sent a bill to the Scranton city council in January 1919 demanding reimbursement for losses of $644.79 suffered by the company during the suspension of service. The mayor of Scranton refused Murphy's demand, saying the L&WV police should have kept the boy out and thereby prevented his death. Undeterred, Murphy persisted and asked the mayor to submit the bill to the council. In March, the city officially rejected the claim on the advice of its solicitor.

The World War Affects Operations. While the power company was experiencing service disruptions and undergoing a major rebuilding, the Laurel Line continued to operate with its normal 20-minute headways. In a representative period of two months during the spring of 1918, an average of 75 cars per month experienced delays. Reasons ranged from loading papers, unloading baggage, failures of brake rigging, air hoses, couplers, and circuit breakers, to fires on cars, hitting pedestrians, and a cow on the track. The most common cause of delay was a failure of power for various reasons. On only seven days per month did all trains run without interruption.

The entrance of the United States into the World War in April of 1917 resulted in government pressure on the railroads to consolidate service and save fuel and manpower. In May, Daniel Willard, the president of the Baltimore and Ohio Railroad, who was also serving as chairman and transportation expert of the Advisory Commission of the Council of National Defense, urged the steam railroads to discontinue as far as possible their unnecessary and competitive passenger services.

Frank Ford, chairman of the executive committee of the L&WV, was quick to see an opportunity for the electric line in this situation. He promptly called on L.F. Loree, president of the Delaware and Hudson Railroad and also chairman of the Committee on National Defense of the American Railway Association for the Eastern Department. He suggested the D&H could discontinue practically all of its passenger service between Scranton and Wilkes-Barre, which the Laurel Line could then easily handle. He even resurrected the old dream

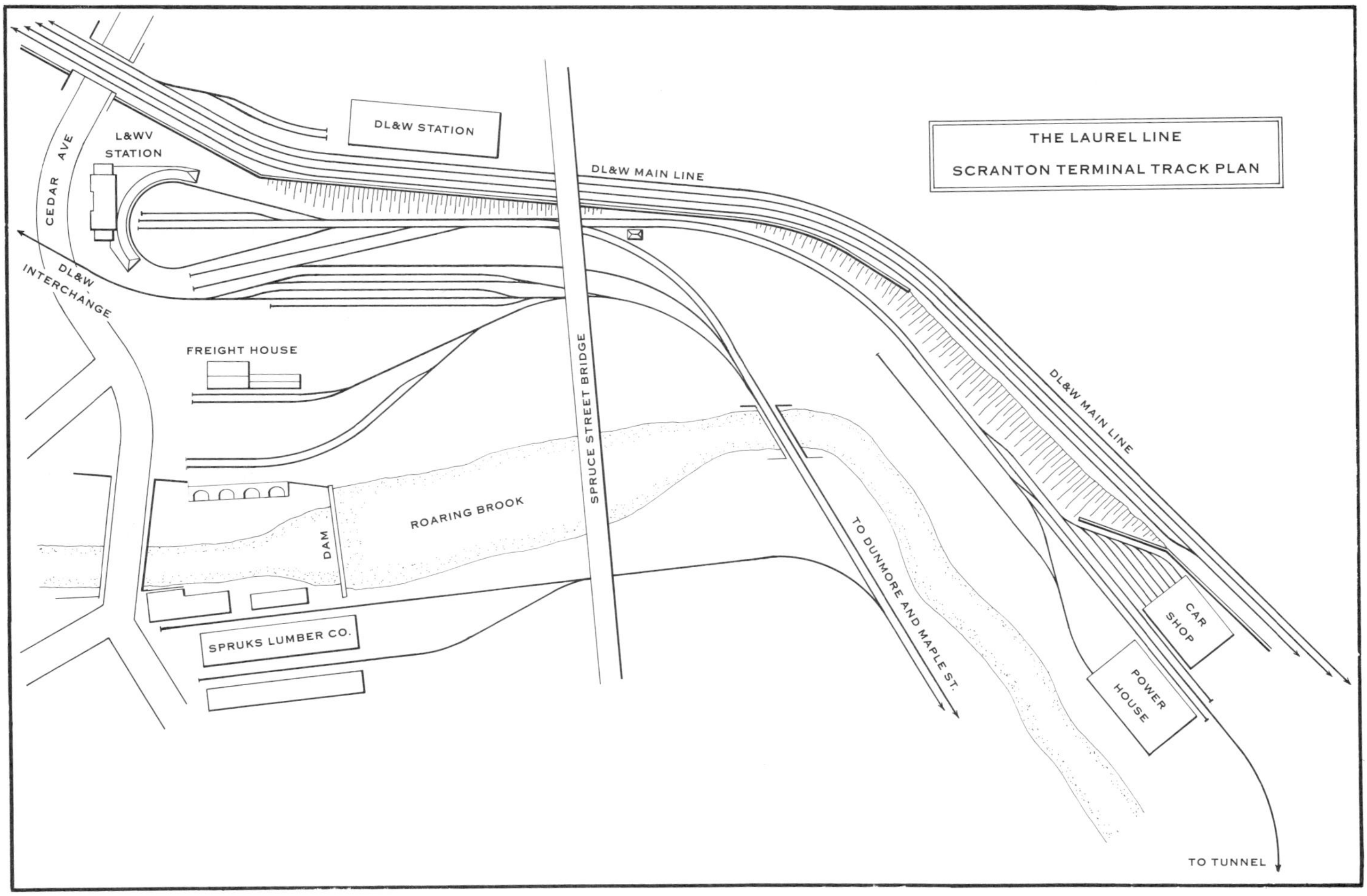

At the end of the roaring twenties, the Scranton terminal had a well-established look to it. A two-man wooden car enters the terminal loop in the right background, possibly in Dunmore service. The excursion train shed, to the left of the station, could still be used. A locomotive busily drills freight cars in the yard. And, at the far left, an express motor rests near the freight station.
Edward S. Miller Collection

Prior to World War I, Pittston was an expanding city, its hills peppered with two-story frame houses. The population exceeded 18,000 in 1920, more than twice the number living there today. As the most important place between Scranton and Wilkes-Barre, it rated both freight (near) and passenger (far) stations. The freight station is now used by the city as a storage building adjacent to the community swimming pool. Because of "slow" films and plates of that era, the southbound wooden passenger car was probably not moving into the station but was posed just short of the telephone pole on this spring morning. The public team tracks contain three box cars to be emptied by local consignees or loaded by local shippers. The third track beyond the passenger car is the Erie Railroad's branch into Pittston.
Alan Martlew Collection

of going to Carbondale by electrifying either the D&H or the New York, Ontario and Western.

Loree was cool to these proposals, pointing out that the public would make loud complaints and the Public Service Commission might not approve. Ford pointed to a map and noted the L&WV stations were all within a half mile of the major D&H stations. The few exceptions could be served by existing streetcars or new motor bus lines. Ford informed Murphy of this conference and urged him to see what he could do locally.

Murphy quickly discovered that the local officials of both the D&H and the Central Railroad of New Jersey had no power to change schedules. At the time, the D&H was operating 10 trains in each direction between Scranton and Wilkes-Barre and the Jersey Central was running three trains each way. The only concrete action taken by the D&H was the removal, in June, of two trains each way from its schedules. Thus the pressure of the war did not result in the Laurel Line getting its hoped-for monopoly of service in the area.

The rapid growth of wartime traffic overwhelmed the railroads' ability to handle it efficiently. Consequently, on December 26, 1917, Woodrow Wilson proclaimed the nationalization of the nation's railroads and the formation of the United States Railway Administration (USRA), which was placed under the control of the Secretary of the Treasury, William G. McAdoo, who then assumed the title of director general of the railroads. Depending on the circumstances, the L&WV had developed the practice of either stressing that it was the equivalent of a steam railroad, which just happened to be operated by electricity, or that it was an electric interurban railroad heavily dependent upon short-haul passenger traffic. In this case, the L&WV took the latter stance, which was officially recognized by the USRA and therefore the line was not nationalized.

Nonetheless, the road cooperated with the federal agency. In response to an inquiry from the assistant director general asking what reductions had been made, Murphy claimed the L&WV was ". . . principally a passenger road." He noted that 80 percent of the gross earnings came from passenger traffic and he claimed the line handled between 97 and 98 percent of the total passenger business between its terminal cities. In addition to the slight reduction made by the D&H, the Lehigh Valley removed one train each way between Pittston and Wilkes-Barre and the Lackawanna eliminated a round trip between Scranton and Kingston. Murphy said the L&WV was still operating on a 20-minute headway with single cars running until 2:00 P.M., after which two-car trains ran until midnight. He noted that increasing the headway from 20 to 30 minutes would save only two men and since the line had its own culm piles, it was not in the market for coal. The L&WV service remained essentially unchanged, despite the war. However, on September 15, 1918, a 30-minute headway was established for midday service.

Although there probably was not much real danger, Murphy became alarmed at the beginning of the war about possible sabotage. He wrote to Major Vail of the 13th Regiment of the Pennsylvania National Guard requesting troops to patrol the Avoca and Hollenback viaducts, as well as the Scranton tunnel, to protect them from ". . . possible destruction by cranks and German sympathizers." Evidently, the military authorities agreed with Murphy, for a unit was sent to guard the structures. By October, calm returned and the unit was withdrawn.

To a certain extent, the war increased passenger loads. Restrictions on automobile travel on Sundays in 1918 resulted in higher passenger earnings. After the war ended in November 1918, there was a brief spurt in traffic as soldiers and munitions workers returned to the area. Partly as a byproduct of wartime inflation, fares gradually increased throughout the period. In 1914, the one-way fare between Scranton and Wilkes-Barre rose from 30 to 35 cents and by the end of the war had jumped to 50 cents. There were proportionate increases in the round-trip rate until January 1918 when all round-trip fares were canceled. Commutation tickets were introduced in 1917 at a rate of ¼ cents per mile, good for 30 days. The mileage rate rose to two cents per mile by July of 1918. These tickets supplemented the mileage books, which were available in denominations of 200 or 400 miles at the same basic rate. In October 1918, minimum one-way fares were increased from five cents to seven cents in line with similar increases on the local street railways in both Scranton and Wilkes-Barre.

Supplementary income was gained by leasing space for newsstands in the stations at Scranton, Pittston and Wilkes-Barre. The monthly rental amounted to $300. Oddly enough, Murphy rejected the use of power-saver recorders on the cars, which were becoming popular among electric railways of the period, as a cost-saving device. His reasoning was that all of the passenger cars had automatic controls and little saving could be anticipated. Since the company was paying only 75 cents a ton for its fuel, the savings would be negligible.

Murphy frequently had to deal with minor complaints. One centered on the cigar and newsstand in the Pittston station. C.C. Bowman, an ex-state Representative from Luzerne County, wrote several letters, including one to Ford, complaining of the clouds of tobacco smoke that passed through the swinging door separating the shop from the waiting room. "As there is only one waiting room at the station, ladies and other patrons, some of whom have affections of the air passages and lungs, suffer much distress." Murphy dismissed Bowman as a "crank" with a quarrelsome disposition and said the only way to stop smoking at Pittston would be to hire a policeman, which the company could not afford. In typical fashion, Murphy complained that the Pittston population was ". . . difficult to handle. There are a good number of foreigners who have very little consideration for the sensibilities of men like Mr. Bowman. . . ." He argued that if they closed the door completely, as Bowman wanted, they might as well close the cigar stand and lose the $75 monthly rental.

An old problem which had plagued Charles Conn continued to annoy his successor. Partly as a result of the closing of theaters and other public places due to the flu epidemic, Murphy received many complaints of young men congregating in the stations or on the cars and expectorating on the floors. The vice president wrote to the Department of Public Health to have the police stop the congregating and to post notices prohibiting expectorating. A health official informed Murphy that under Pennsylvania law the railroad itself could arrest loiterers with its own special officers. He pointed out the same law required the railroads to post bills prohibiting spitting. Under the statute, if they failed to do so, the carriers were subject to a fine.

In October 1919, Murphy received an eight-page letter from a woman in Scranton purporting to speak for several other ladies complaining about the dirt, dust and filth ". . . sticking fast to the floor with a generous sprinkling of expectoration" in the Laurel Line coaches. Describing herself as "an indignant spokeswoman," she was particularly irate about the spitting in the cars and threatened to write letters to the local newspapers and board of health detailing these conditions unless improvements were made. Murphy responded that he shared her desire for cleanliness, but he asserted it was almost impossible to stop the expectorating, ". . . due to the class of public traveling." He encouraged her to write a ". . . brief letter and mention an explicit case where you find cars in a bad condition. . . . I would like to have this letter so I can post it on the trainmen's bulletin board." Whether this method of encouraging the crews to enforce rules of cleanliness was successful is impossible to say.

Even trivial matters were often blown out of proportion. For instance, in response to a complaint from a priest in Inkerman about the conditions in the station there, Murphy again blamed the traveling public and said the company could not make people behave. He attributed delays in service to the effects of the war, saying parts were very difficult to obtain and labor was not doing conscientious work because the men were indifferent to the company and the public.

An unusual dispute over a five-cent fare ended in an appeal to the courts. A passenger traveling from Scranton to Wilkes-Barre claimed he had to leave the car at Hancock, receiving a slip from the conductor when he did so for use on the following car. However, the officious conductor on that car refused to accept it and demanded an additional five-cent fare. When the passenger did not immediately get reimbursed, he filed a lawsuit for the amount. Needless to say, the case was quickly settled out of court.

Park Traffic Declines. During the decade of World War I, the amusement parks along the line continued to provide some business, but on the whole travel was down. Valley View Park, located in Inkerman, which was owned by the L&WV, continued to function as a picnic grove. In an effort to revitalize this location, the company made an agreement with A. Ogden of Philadelphia in April of 1915, whereby he would erect amusements such as a merry-go-round, shooting gallery, penny arcade, doll baby rock and a novelty stand. For this, he would pay 10 percent of the gross receipts from the amusements to the railroad company. In keeping with past

At Valley View Park station, Market Street, Inkerman, a southbound train pauses about 1908, its second car partly obscured by the waiting station. The park is on the far side of the track.
G. Dewey Shrader, Edward S. Miller Collection

practices, no intoxicating beverages were allowed. Despite this attempt to modernize the park, it continued to falter.

Luna Park, in Scranton, which already had amusements, was functioning only on a minimal level. After its initial success in 1906 and 1907, the only paying activity was the dancing pavilion, although the amusement rides were still maintained. All of this came to a fiery end on the night of August 22, 1916, when most of the property burned down. From the Laurel Line's point of view, the sudden demise had little effect on traffic. The park had been very disappointing almost from the beginning and the company never recouped its investment in the extensive station facilities it built there.

The one remaining park along the line, Rocky Glen, was also experiencing difficulties. For a brief period, however, the park's future looked bright when the Metro Film Company considered acquiring the pavilion for making motion pictures. Nothing came of this, but two years later the Federal Feature Film Corporation of New York contemplated purchasing the park for the same purpose and for operating it as an amusement center as well. Evidently, the price of $65,000 was too high, for the company never consummated the deal. Probably a more important reason was the general shift in the industry from the East Coast to Southern California.

In the meantime, a dispute had developed among the owners of Rocky Glen with the result that the property was divided between E.S. Williams on one side and Arthur Frothingham and Miss Annette Reynolds on the other. Both sides claimed that the Laurel Line station was on their respective properties. Due to this dispute, the park had not been operated since 1912, except for swimming and ice skating.

A side issue involving the L&WV centered on an overhead bridge planned for a road a short distance south of the station. Because of high construction costs, the company built the bridge north of the station without getting permission from Frothingham to change the route and cross his property. Early in 1919, Frothingham and Miss Reynolds called at the New York office of Ford, Bacon and Davis to discuss the park and its relationship with the L&WV. Since Ford was out of town, they spoke to George Bacon. They complained that the L&WV had always opposed their plans and cited as an example their attempt to have a local impresario furnish music for dances. They asserted that Murphy had discouraged the plan by stating the L&WV could not unload passengers at the station unless it received permission from Williams, the other litigant, to cross his property to get to the dancehall.

For the forthcoming season, Frothingham announced that he again had made arrangements to open the ballroom and wanted to know whether the L&WV would sell tickets to passengers desiring to go to Rocky Glen. He offered to sell his and Miss Reynolds' interests in the park for $35,000. He also pointed out that the borough of Moosic had never accepted the revised road route and would probably demand that the park and railroad remove the present overhead crossing and build a new bridge on the site of the original road at a cost of about $20,000. Bacon assured Frothingham the company would be most happy to transport passengers to the Rocky Glen station but that it was not interested in purchasing any of the park property.

Somewhat encouraged by the conference, the Frothingham interests made a few additions and opened the park for the season. The Laurel Line was happy to carry passengers there but refused to sell admission tickets to the park. The passenger load was such that Murphy and Ford considered discontinuing service but decided to leave the station open until it needed extensive repairs. A similar situation prevailed in 1919 when the station provided total earnings of about $2,000. In December, the rotting wooden platforms between the tracks were removed and replaced with cinders. Murphy again refused to buy the park on the grounds that it should be operated by park specialists, not railroad people. He told Ford the picnic park at Valley View was all they could handle. Within a few years, however, he was to change his mind.

Disappointing Returns from Mail and Express. During this period, the L&WV again considered the prospects of securing a mail contract. Early in 1913, Conn urged William Higgins to explore this possibility with the Post Office Department. Higgins replied in February that the L&WV could not meet government regulations because it lacked the proper equipment such as water coolers and wash bowls that the post office required. He argued it was not worth the expense of remodeling the cars for the small amount of mail which would be carried.

When the new management took control in 1914, it explored the question again. Ford urged Higgins to seek a post office contract. Higgins reluctantly agreed but he advised Ford that the company could expect to earn only $800 per year revenue and it would be obliged to carry the mail from the station to the post office in Pittston. He feared they would lose their newspaper business which grossed about $900 per year if the mail were carried. After reading this, Ford agreed to drop the matter. As with Murphy and Rocky Glen, Ford would change his position in a few years.

The express business would also prove to be disappointing during the wartime era. The Laurel Line until 1917 was averaging about $15,000 per year under terms of its contract with the Adams Express Company. In May 1917, Adams cut the payment to the L&WV from 43 percent of receipts to 40 percent. Murphy immediately complained, pointing out the DL&W received 45 percent and the Pennsylvania Railroad 50 percent. There was a vast difference between the Pennsylvania or DL&W and the little Laurel Line, as an official of Adams Express pointed out when he rejected Murphy's protest.

One of the steps which the United States Railway Administration took when it assumed control of most of the nation's railroads was to force a consolidation of the express agencies under the American Railway Express Company. This move had an adverse effect on the Laurel Line, which promptly lost about one-half of its business. For example, a manufacturer in Wilkes-Barre formerly sent a car of silk and other materials from that city to Passaic, New Jersey, via the L&WV and DL&W. The new express company rerouted this car via the Lehigh Valley, but the firms involved were not satisfied with the service and switched to trucks. Murphy tried to get the American Railway Association and the American Electric Railway Association to use their influence to protect short lines from traffic diversions, but in this he had little success.

By 1920, the Laurel Line was averaging only $350 a month from express. An occasional car was loaded at the Scranton Transfer station of the American Railway Express Company and routed DL&W, L&WV and Lehigh Valley to the consolidated warehouse of the Central Railroad of New Jersey in Wilkes-Barre. When the American Railway Express sent a new contract to the L&WV, Murphy rejected it. He told F.E. Welsh, the superintendent, the contract was of no value since it provided no guarantees of minimum business. He rebuked the express company for eliminating competition and complained, somewhat unfairly, that the L&WV had purchased four cars at a cost of $36,000, specifically for this purpose. He also pointed out the company had spent $5,000 for a freight house in Wilkes-Barre, especially for the express business, which was now no longer needed. He concluded by ordering the express company out of the line's Scranton freight house, which it had been using for storage. Officially, the last day that American Railway Express operated over the Laurel Line was August 31, 1920.

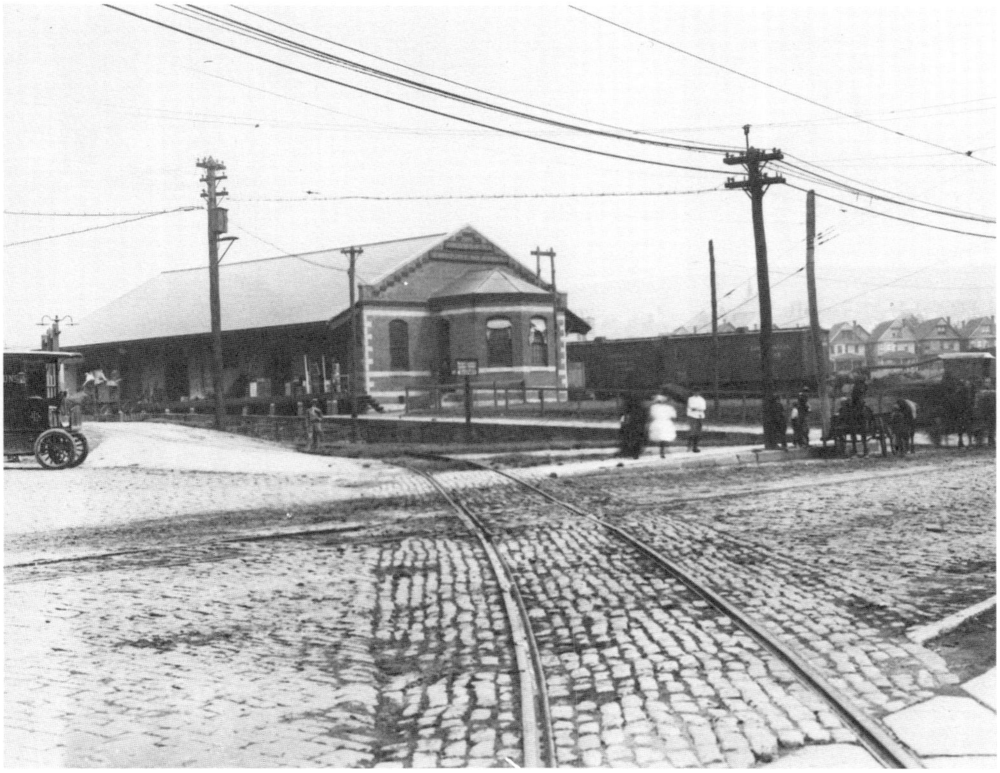

This very substantial building on Cedar Avenue in Scranton served as the L&WV freight station. Railroad freight stations were staffed with an agent, who solicited traffic and supervised other staff; clerks, who were experts at preparing weighbills and rating shipments; and freight handlers who transferred less-than-carload shipments from drays or trucks to railroad boxcars or vice versa. The nationwide abrogation of LCL freight to truck lines, centralized sales forces and computerized billing offices made the local freight station obsolete on most railroads in the 1960s. What transport museum would not covet that battery-powered truck on the left?

George Arents Research Library, Syracuse University

Frustrations with Freight. Freight tonnage, on the whole, did not measure up to expectations. During the first six months of 1914, the L&WV originated roughly 68,000 tons of freight and received about 69,000 tons for a total of 137,000 tons hauled. Of this, agricultural products accounted for about 4 percent; animal and meat products, 2 percent; mine products, especially anthracite, 40 percent; lumber, 7 percent; manufactured goods, particularly oils, cement and alcoholic beverages, 17 percent; merchandise, 19 percent; and miscellaneous items, 11 percent.

As with the question of mail, the new management thought it could make improvements in the freight service. Writing from New York, Ford urged Higgins to investigate the possibility of establishing a freight interchange with the Pennsylvania Railroad at the point where the tracks of the Wilkes-Barre Connecting Railroad crossed over the Laurel Line on its approach to the bridge over the Susquehanna River. Ford claimed about 25 cars a day could be routed over the L&WV to Scranton if such a connection existed. He said he knew some high PRR officials and would like to bring the matter up. Higgins replied that the track would be on a 14 degree curve and a considerable grade. If Ford did pursue the matter further, he received no help from the Pennsylvania, which was traditionally hostile to electric roads in any event.

Nor was the Lehigh Valley any friendlier. Murphy complained in 1915 that he had submitted a bid to the Lehigh Valley Coal Company to haul coal from the Broadwell property to the Heidelberg Breaker for 11¢ a ton, which was a low price. Instead, the contract was awarded to the Delaware and Hudson and Lehigh Valley Railroads, which Murphy felt could not handle it any cheaper. Obviously, the coal companies favored the routes of their parent railroads whenever possible.

In an effort to increase freight traffic, the line prepared another industrial folder, less elaborate than its first one, showing factory sites along the right-of-way between Scranton and Rocky Glen. Many copies were distributed in 1917, but nothing was gained and no new companies located along the line.

As it was, handling existing traffic presented sufficient challenges to management. Typical was a letter from Gaynor Contracting Company in 1916, complaining about an incoming carload of brick, which was placed on the interchange track with the Erie Railroad in Dunmore at 12:00 noon. The contracting company called the Laurel Line office to determine when the car would arrive at the terminal freight station, located just two and one-half miles from Dunmore. Assured by the freight office that the car would be down

By the end of its first decade of service, the line's newest locomotive, number 403, looked rather shabby as it wheeled nine tank cars of oil off the DL&W interchange track in Scranton. The photo is not precisely dated but it was after 1926 when the sixth floor was added to the DL&W passenger station and before 1934 when Edgar Boland was hired, and who later claimed that none of the locomotives had bells on the roofs when he started working for the L&WV. The engineer is Joseph A. Smith of Pittston, who had worked for the line since 1904. Resting on the front platform of the engine is a pole, which could be held into the poling pockets in front of the engine and thus used to pole or move freight cars on an adjacent track. *Edward S. Miller Collection*

One of the early freight customers was the ramshackle brick plant (ABOVE), along the northbound track at North Main Street and Saylor Avenue, Plains Township. The facility last shipped brick via the L&WV in 1922. Plans were drafted late in 1911 to relocate, straighten, and raise the tracks at the brick plant. *Edgar J. Boland, Edward S. Miller Collection* **The new fill (RIGHT) was in place for several years before the main line tracks were moved onto it. One of the old main line tracks remained to serve the plant; it was later used for coal loading into the diesel era.** *Lackawanna Historical Society*

within the hour, the contractor ordered six teams of horses and a number of men to be ready to unload it as soon as it arrived. After waiting more than an hour, they talked to an official at 2:00 P.M. who informed them, ". . . that the car [is] on the way down." By 4:30 P.M., the car still had not arrived. Again, Laurel Line officials claimed a crew was after the car and it would be there shortly. Finally, at 5:30 P.M., the tardy car made its appearance. Gaynor was justifiably critical of the L&WV performance, since he and his men wasted time and money. He informed Murphy, "If this is the kind of service we are to expect, it certainly will be very much better to haul our materials direct from the Erie Station.

While poor dispatching and a less than honest approach to the customer was responsible for that fiasco, sloppy performance by crew members caused more trouble. For example, in 1919 a freight crew servicing the Pennsylvania Coal Company siding attached a cable to an empty car, using a utility pole as a pivot, to pull it onto the main track. They managed to move the car but in the process they almost pulled down the pole and its attached power lines. William Jennings, the general superintendent of the coal company, demanded that Murphy investigate and prevent a recurrence.

The vice president had other questions confronting him. The company had sublet a gravel bank near Moosic to a local contractor who sold the gravel at cost to the L&WV. Murphy discovered the cost per ton for this sand and gravel was almost one dollar, yet the same material could be purchased from outside sources for only 30¢ a ton, plus 28¢ delivery charges, or a total of 58¢ a ton. He told H.P. Decker, the contractor, if he could not reduce his cost, the L&WV would make other arrangements. Five weeks later, the men were laid off and the operation was shut down. Thus the L&WV's poor record in handling its own gravel, as illustrated earlier by the Meadowbrook crushed stone operation, continued.

A positive accomplishment was a joint agreement reached by all the railroads in Wilkes-Barre in 1918 for the inspection and repair of cars interchanged there. The railroads, which included the Pennsylvania, Lehigh Valley, Jersey Central, Delaware and Hudson and the L&WV, would share the crew which consisted of a foreman, two clerks, and 18 inspectors and repairmen. Each company was assessed an amount based upon the number of cars it interchanged per month. This agreement benefited all the carriers and reduced labor costs.

This attractive brochure was issued in 1906 in an effort to gain more freight business. Printed in color, it was much more stunning than the 1917 release. *Edward S. Miller Collection*

Paternalism in Personnel Relations. Such agreements were beneficial to the L&WV because wages rose rapidly during the wartime era. In 1914, motormen and conductors earned 32½¢ per hour the first year and 33½¢ per hour succeeding years. When in charge of trains handled by electric locomotives, an additional 5¢ per hour was paid. Brakemen received 26½¢ per hour and switchmen 22¢ per hour. By 1918, motormen and conductors were getting 42½¢ per hour, and on electric engines 46½¢. Brakemen earned 35¢ and switchmen 30½¢. As of September 1920, motormen and conductors got 60¢ per hour, on electric engines 63¢ and brakemen 55¢. Other employees were paid proportionately. For example, in 1916 freight clerks earned from $27.50 to $55.00 per month, trackmen 17¢ per hour and gate watchmen 12¢. Fares had to be increased periodically to cover these additional payroll costs.

A statistical analysis of employees in 1917 showed there were seven general officers, earning $13,920; six male clerks, who received $7,200, and four female clerks receiving $2,400. The maintenance forces included 120 men who earned $41,000 and transportation numbered 76 men, who were paid $83,810 for the year. Hours worked per day ranged from eight for clerks, firemen and policemen, to nine for carpenters, conductors, engineers, flagmen and agents, to 10 for blacksmiths, brakemen, inspectors and track workers. Not counting the office help or administration, the line employed an average of 360 men per month.

There were no serious problems with labor organizations during this period. However, there was a flurry of excitement in 1914 when some workers attempted to organize a new union. The company responded with a policy of retrenchment. Although the central labor union protested, no further action was taken. In 1918, the men voted unanimously to strike unless they received a pay increase in accordance with that recommended by the Railroad Wage Commission for the federally controlled roads. After the general council of the United States Railroad Administration publicly asserted that the L&WV was not under government control, the men settled for a somewhat lower increase and a strike was averted.

The weaknesses of the labor unions were reflected by Murphy's paternalistic attitude toward his employees. He warned trainmen in 1915 that they ". . . must discontinue sitting around the public waiting room at the terminals talking to ladies, as has seemed to be the practice of late." In March 1916, he received a letter from a local grocer complaining that one of the L&WV workers owed him more than $23 and refused to pay. Murphy told the grocer he had requested the employee to make some effort to pay his bill, adding, ". . . in future, you will please take matters of this nature up with the customer as it is not our practice to devote attention to the collecting of bills contracted by employees."

What he did not say was that on the same day he wrote to the worker's wife, chastising her for her poor management of the household, since the worker claimed he turned his earnings over to her. Murphy concluded, "You will kindly make some effort to pay this bill, otherwise, we shall be obliged do dispense with the services of your husband." No doubt there was a lively discussion of this matter that evening in the employee's household.

World War I caused a severe drain of manpower from the Scranton area. All industries, including the railroads, experienced labor shortages. In the case of the Laurel Line, Murphy reported there was no problem with the officials "or men in important positions," as they had families or were over the age limit for military service. Between the draft and other lines paying higher wages, the company found it hard to retain skilled labor with a consequent decline in maintenance and repair capabilities.

The wartime inflation made it difficult to attract capable workers, and by 1919 the line needed both a master mechanic and a chief engineer for the power plant. Women were hired as clerks in the Scranton and Wilkes-Barre freight stations and performed their tasks satisfactorily. As the war came to an end, former employees drifted back and some even requested special letters in order to obtain an earlier discharge.

Right-of-Way. Routine maintenance continued along the right-of-way during the decade. An analysis of the property in 1914 revealed there were 13 stone bridges and 32 iron bridges, which together had a combined length of 2,786 feet. Thirteen highway bridges and two trestles crossed the track as did eight railroad bridges and trestles. On the 22.63 route miles, there were 77 curves totaling 11.67 miles. Only 1.54 miles of line were on level ground.

A few minor changes along the line were made and expansion was briefly considered. In 1913, the proposal to build an interurban to Lake Ariel, which had been discussed in 1909, reappeared. This time Gerald MacDonald, president of the MacDonald Construction Company, proposed to build a line from Dunmore to Lake Ariel at an estimated cost of $250,000. He wanted to use the Laurel Line tracks from Dunmore to the Scranton terminal as well as the carhouse and power facilities. Calvert Townley, who was still in control at that time, encouraged MacDonald but made no commitments. After extensive discussion, the L&WV offered MacDonald a 25-year contract, whereby the Laurel Line would handle the Lake Ariel interurbans for a fee of 75¢ per car in each direction and perform ticket and baggage sales as well. The company also offered a portion of its unused right-of-way beyond Dunmore for $500 annual rental.

As happened in the earlier scheme, MacDonald was unable to secure the necessary financing and the plan was shelved. Such a line could never have succeeded at this late date since the sparse population, combined with the development of the automobile, would have precluded any significant traffic. Three years later, officials of Jessup Borough, about six miles north of Scranton, tried to entice the L&WV to extend its tracks from Dunmore to that town to carry coal then handled by the D&H and DL&W. Murphy promptly rejected this suggestion as being too expensive and unnecessary. Thus, the last flickering dream of expanding the Laurel Line beyond its existing limits died out.

A few changes were made in sidings to plants. In 1915, agreement was reached with the Erie Railroad for the construction of a siding from the Dunmore line to a nearby washery. Three years later, a new siding was built in Wilkes-Barre to serve the Wilson and Company facility, which was leased to them for $600 per year for 10 years with an option for renewal for an additional 10 years. This firm was to provide a considerable amount of business to the L&WV for much of its life. On the other hand, the stock runway located at the Adams Express depot in Wilkes-Barre was removed in 1917, since it was seldom used and the space was badly needed for cargo loading.

Another improvement made at this time was the construction of a dispatcher's building at Scranton Junction costing about $500. When it was completed the dispatcher moved there from his previous quarters in the terminal station. Not approved by the consulting engineers was a proposal by the Wilkes-Barre Chamber of Commerce that the ". . . unsightly and out-of-harmony" wooden trolley poles in the city be replaced with ornamental iron ones. The L&WV favored beautification only if it were free.

A workman shovels silt, deposited by run-off from the Butler Colliery above the North Pittston station. Although a source of revenue, coal mining and processing produced never-ending problems for the L&WV.
Edward S. Miller Collection

A recurring source of trouble for the company was the robbing of coal from veins located under the right-of-way. It was estimated that under the Scranton terminal area alone there were 1,471,680 tons of coal. This did not affect the company, since it had an agreement which precluded taking coal from those veins. But many areas under the tracks were subject to exploitation by the mining companies. In 1913, the Lehigh Valley Coal Company reported it was about to rob the coal in the Clark Vein near Heidelberg. It offered to sell four pillars containing 365 tons of coal to the L&WV at $1.25 a ton. For various reasons the Laurel Line management declined the offer and instead posted a watchman and issued slow orders for that section of the line. Eventually, the southbound track was affected until fill could be applied, thus stabilizing the surface. Other diggings under the tracks at Moosic and Avoca caused only minor delays.

The coal robbing and attendant adjustments were costly to the L&WV. In some cases, the tracks had to be moved temporarily while the robbing was underway. This was the situation at the Butler Colliery when the Pennsylvania Coal Company began robbing operations in 1919. As the mining dragged on, Murphy wrote to W.P. Jennings, the superintendent, complaining that the operation had already cost the L&WV $1,000 plus additional expenses for watchmen at the site. He told Jennings that ". . . there should be more cooperation so as to give this company and the public better service. We have now several slow orders on your operation, some of which have been on for a long time."

On the other hand, coal represented revenue for the L&WV, since a considerable part of its freight traffic consisted of hauling anthracite. There were several breakers located along the line and the company looked for opportunities to carry coal from off-line mines to these facilities in order to increase business. Sometimes, trouble resulted. In 1916, the L&WV opened negotiations with the Erie Railroad for the

The tunnel-lining car is loaded with sand and gravel from bins erected above the north portal in 1916. Note that the engineers have penetrated the tunnel portal to permit a smooth flow of materials through the chutes. The "Over-the-Hill" line crossed the top of the embankment and permitted easy filling of the bins.
Edward S. Miller Collection

purpose of bringing in anthracite to be processed at the Minooka Breaker. Before the deal could be completed, the Minooka Coal Company failed and was later sold to Spruks Brothers for $37,155. Included in the sale was the siding from the L&WV main line. Spruks rented the breaker to another party who made his own arrangements for procuring coal. As a result, the plan to handle coal from the Erie Railroad collapsed.

A major project during this period was the lining of the Scranton tunnel. Evidently, problems had developed with the roof of the tunnel, which, except for a few sections of solid rock, was covered mostly with timber. The concrete lining, which was applied to about two-thirds of the length of the tunnel, would eliminate the dangers of falling rock and rotting timber. In February 1916, the board of directors approved the project at an estimated cost of $87,000.

To facilitate the work, the engineers designed a tunnel-lining car, which was built on the frame of an old car acquired by Ford, Bacon and Davis which had been damaged in a fire. Mounted on the car was a pneumatic placing machine; cement, sand and gravel bins; eight-inch steel and iron pipe, with a section of six-foot flexible rubber hose; and two 750-gallon water tanks. A crew of eight men plus a foreman and compressor engineer did the work, averaging about 70 cubic yards placed each eight-hour working day. The work was done in the summers of 1916 and 1917. During the work hours, scheduled passenger trains were rerouted on the over-the-hill line.

Following completion of the project, the company offered the tunnel-lining car for sale. Ford, Bacon and Davis contacted Westinghouse, Church, Kerr and other prospective buyers, but no interest was expressed in the car. Murphy then suggested the car could be stripped of its equipment and converted into a work car, using the truck of another car, which had been damaged in an accident. He proposed putting a derrick on one end for use in loading rails and other mate-

Workers demonstrate the use of the car designed to facilitate lining the Crown Ave. tunnel in 1916. The car is on a siding opposite the car house.
Charles Sobeck, Edward S. Miller Collection

rials. The cost of labor was so high and of such poor quality, according to Murphy, that the old methods were no longer efficient or safe. Up to this point, the line had been borrowing equipment, primarily from the Erie, when heavy work or wreck repairs were necessary. Ford did not agree with Murphy and the car remained for sale.

Minor Aggravations. On occasion, the penny-pinching practices of the Laurel Line management caused the company to expend even more money. When Cedar Avenue in front of the Scranton Terminal was repaired by the city, Murphy refused to pay the assessment of $1,360, claiming that his company's responsibility for paving had been met by the Lackawanna Iron and Steel Company's original surfacing of the thoroughfare with cobblestones. In contrast, the DL&W, which owned the land on the opposite side of the street, agreed with the city's position. The matter was taken to court in 1914 and ultimately a court-appointed referee decided against the company. As a result, the line had to pay the original assessment plus $272 in interest and costs.

Other towns along the line managed to get into disputes with the company. The borough of Avoca demanded a guard be placed at the South Street crossing in 1915. Murphy claimed a watchman was unnecessary since all of the local trains stopped at the station; the limiteds, which did not stop, whistled for the crossing; and as there were two other crossings, one an overpass and one an underpass, South Street was used mostly by pedestrians. The borough threatened to go to court; records do not indicate how the matter was resolved.

The volunteer fire company of the neighboring community of Dupont solicited a contribution from the line in 1919 in order to purchase more hose and a new fire truck. The chairman of the volunteers reminded Murphy that when the Dupont station caught fire the year before, they had extinguished the blaze even though they had to form a bucket brigade because of the absence of a standpipe and the lack of hose. Murphy replied by saying, "It is not customary for the company to contribute to fire companies and matters of this sort and I do not find that the Interstate Commerce Commission allows such contributions, the assumption being that taxes should take care of such matters." To take the sting out of this rejection, he enclosed his personal check for $15.

The Laurel Line Updates Its Equipment. Maintaining and updating the line's equipment was an endless task. A long dispute erupted with the Baldwin Locomotive Works in 1913 over the failure of the elliptic springs on car number 11. The L&WV claimed the springs were defective and demanded their replacement without charge, while Baldwin blamed rough treatment and shock from the roadbed. Since car 11 was the only one affected, Baldwin was eventually forced to agree and made the necessary replacements. All of the cars were supplied with a new type of air equipment in 1915. While the adjustment was being made, switchmen were warned not to couple cars with different types of air as they were not interchangeable. In a deviation from company tradition, the motormen were invited to criticize the operation of the new brakes.

Type M car 13 models the simplified paint scheme inaugurated in 1915. The car was equipped with Westinghouse 251-C switch group; such apparatus was retrofitted to these cars in 1920-1921 when the new 557 motors were also installed. However, the trucks are the original 6'-0" wheelbase, curved-equalizer design from Baldwin that were apparently later supplanted by a longer wheelbase truck. All of the apparatus then on this car was transferred to one of the new steel cars in 1928; the body of 13 remained until 1931.

George Arents Research Library, Syracuse University

A new system of coach painting was begun in 1915. Where badly worn spots existed, the wood was repaired, puttied and given coats of primer. The entire body was sandpapered or smoothed with a fine carborundum cloth. A coat of "surfacer" was applied and the following day a coat of color. The next day the car was striped and lettered. A gold stripe, one-half inch in width, was placed two inches above the bottom of the siding. The numbers were in four-inch numerals. The lettering LAUREL LINE, also four inches, was placed in the center of the car over the windows and three-inch letters designating L&WV were applied over the first and last windows. Roman-style letters and numerals edged in black were used. Two coats of varnish were then put down to protect the surface. Interiors were scrubbed, touched up and varnished. The master mechanic reported the first car so painted had been in service for more than two years and was still in good shape. Average cost of this work was $175 per car.

Difficulties, some minor and some major, continued. Considerable excitement resulted when the Public Service Commission of Pennsylvania held hearings on the question of toilets in electric railway cars. For a company which refused to supply even drinking water on its coaches, the problems associated with installing and maintaining toilets seemed monumental. Murphy sent a representative to the Harrisburg hearings who informed the commissioners that the maximum time between terminals was 40 minutes and therefore the Laurel Line should not be required to make this improvement. Evidently the commission agreed, for the L&WV continued to operate in its traditional manner, unencumbered by modern creature comforts.

So many failures were experienced with the motors that in 1916 Murphy requested the Westinghouse Electric and Manufacturing Company to send a representative to study conditions and make recommendations. The Westinghouse engineer identified a number of problems. He discovered the line relays, which opened the control circuit whenever power was interrupted, as in sleet storms or gaps in the third rail, had been cut out for at least four years. As a result, there had been flashing and burnout of motors. He strongly recommended either the line relay be cut in or that a new type high-speed line switch be installed. The line switch, type 26Li-K-3, would automatically open when there was a sudden drop in voltage. Murphy decided to order one for test purposes..

The engineer recommended that carbon discs be substituted for the valuable platinum discs, which were being stolen, in the series line switch, and that a new fitting be installed on the air supply pipe to prevent freezing. In explaining the troubles experienced by the two locomotives, he recommended a larger size lead battery be installed and that the engineers be instructed to operate the machine in series position rather than in parallel when performing switching operations.

He also suggested applying a modified brushholder on the type 50 motors. The report of the Westinghouse representative indicated once again the poor operating practices which had developed on the Laurel Line. This was due partly to the lack of skilled labor during the wartime period, but certainly management had to bear its share of the responsibility for not developing efficient maintenance standards.

More alarming to management was the growing cost of maintaining the motors on the cars. The master mechanic reported that on one day in 1918 he had on the floor a total of 43 armatures that needed rewinding; some required shafts and commutators; and eight had to have cores rebuilt. A total of $26,648 was spent that year simply for motor maintenance. The master mechanic said, "If we had an up-to-date, modern Interpole, Ventilated Motor on this system, our savings on this particular part of the equipment would be at least $20,000."

Careful records were kept on the motor mileage accrued by each car on a monthly basis. For example, in December 1918 only five of the type L cars were in service and they averaged about 3,000 miles. The type M cars averaged about 5,000 miles. All of the combines averaged roughly 5,000 miles. Three express cars ran about 1,000 miles each and work car 301 rolled only 25 miles. The two locomotives ran a total of 3,000 miles.

The continuing problem with the motors on the passenger cars and related maintenance expenses caused concern; an average of six or seven cars were out of service at any one time awaiting repairs. Despite this, when the American Electric Railway Association sent a questionnaire concerning spare cars and equipment for war service in April 1918, Murphy reported the company had seven passenger cars and one combine as spares which could be used elsewhere. Both were type L's built in 1903 and were presently valued at $12,500 each. No doubt the maintenance difficulties with the motors under these cars made the line willing to dispose of them.

By 1919, new motors had become essential. M.F. Flatley, the master mechanic, wrote to Westinghouse Electric and Manufacturing Company inquiring about new equipment. Westinghouse suggested an application of its 557-A8 motors, which, on a car weighing 78,000 pounds, would propel it on 500 volts, using a gear ratio of 17:60. In an emergency, a single motor of this type could propel the car up a 2¼ percent grade one mile long. With modifications, the Baldwin trucks could support these motors.

Murphy decided to order two of these motors complete with gears, pinions, bearing and related switch group equipment at a price of $3,300. Westinghouse agreed to give the line nine months to try them out and promised they could be returned if they were of insufficient capacity or not desirable in the service. The test was successful and in January 1920 the line signed a contract with the Westinghouse Electric and Manufacturing Company for 15 sets of railway motors.

Although Westinghouse shipped the motors within the time specified, there were considerable delays in sending the gears and gear cases. This, and the method of payment stipulated by Westinghouse, evoked a major controversy. In signing the contract, Murphy overlooked the fact that Westinghouse required 50 percent of the bill to be paid at once through a sight draft drawn on a local bank. Murphy had expected to receive a bill directly, rather than make payment to a bank before he could claim the equipment. He wrote an angry letter to Westinghouse complaining about the delays in shipping the gears and cases and the method of payment required. He told the manufacturer, "You are so persistent you have to have your money promptly, but we do not get the material. We should not make any payment on this equipment until it is delivered complete to us for use. All of these delays are your fault."

F.P. Ball, the Westinghouse representative in Philadelphia, tried to mollify Murphy and politely pointed out he should have read the contract before he signed it. He referred to Westinghouse's long interest in the Laurel Line and claimed that ". . . in the past from an accounting standpoint our treatment has been of a very lenient nature."

Murphy was still not satisfied nor mollified and informed Westinghouse he would hold the remaining payments for the additional days of delay until the gears were received. He accused Westinghouse of ". . . either extreme negligence or no intention of making delivery according to contract." The whole affair reflects a rather sorry attitude on both sides. Murphy had reason to be upset over delivery delays which had been characteristic of Westinghouse in the past. The manufacturer, on the other hand, was only too familiar with the tardy payment practices of the L&WV. In any event, Murphy was unwise in striking such a belligerent tone.

Nonetheless, the equipment finally did arrive and was installed on 16 cars at a cost of $80,000. The control equipment on the 10 remaining cars was overhauled. By this time, as will shortly be related, three cars had been sold to the Galveston-Houston Electric Railway and one car had been wrecked. Murphy claimed that the line could operate more efficiently with 26 cars with new equipment than it could with the 30 unmodified cars. He expected only three or four cars to be out of service for repairs at any one time and therefore advised against the purchase of new cars at this point.

Meanwhile, the line felt the need for an additional locomotive. This machine, numbered 403, was built in the Scranton shops between 1917 and 1919. Westinghouse Electric supplied a type HL unit switch control for the four type 50F motors which the L&WV intended to supply itself. The price for the control cable, switches and related apparatus was $3,500. Originally, the company planned to have the Baldwin Locomotive Works supply a new set of trucks, but, as it developed, a fire damaged number 402 in 1920 and its trucks were transferred to the just completed 403. Mechanically, number 403 was similar to number 401. It had four Westinghouse 577A6 motors of 195 horsepower, with a 16-67 gear ratio and a 33-mph balancing speed. It weighed about 52 tons, had a 200 percent overhead, one-hour tractive effort of 41,200 pounds and a continuous rating of 8,800 pounds. Number 402 was eventually rebuilt and received motors and trucks from one of the older passenger cars.

A new lighting system was applied to the cars beginning in June 1919. The old cluster lights in the roof and the side lamps were removed and replaced by 72-watt lamps with shades in the ceiling. Two battery lights were the only fixtures remaining on the walls. Cost of installation was $56 per car and estimated savings per year, based on smaller lamps, amounted to about $12 for each unit. In addition, it was expected that breakage would be much reduced and better illumination was provided. As cars were shopped for painting or major repairs, this change was made. A new type of headlight was installed on cars 111 and 112 for experimental purposes in 1915. Because of its size, it had to be removed when the cars were in trains, as it lacked sufficient clearance. The tests were not that successful and no additional models were ordered.

Construction of locomotive 403 was begun in the company shops in Scranton in 1917 but it is probable that the machine never entered service before 1920. It is shown in 1921 posed with six newly rebuilt side-dump ash cars in the Scranton yard. *George Arents Research Library, Syracuse University*

FOR SALE

High Speed Interurban Passenger Cars

7 Passenger Cars 1 Combination Car

For Immediate Delivery

PASSENGER CAR

COMBINATION CAR

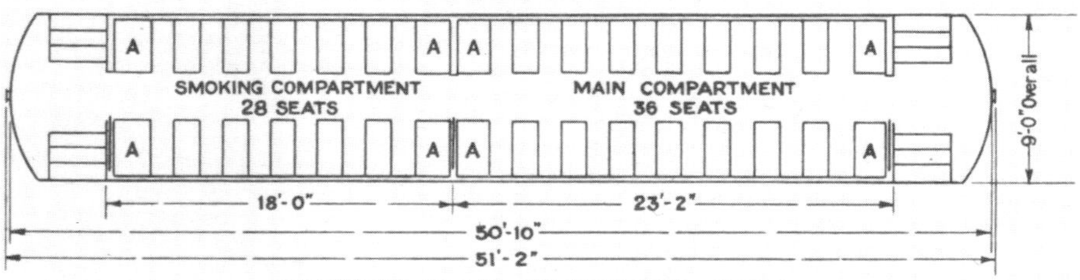

SEATING PLAN—PASSENGER CARS

SPECIFICATIONS PASSENGER CARS

Builder—Amer. Car & Foundry Co.—Type L
Length over body (wood underframe)—50′ 10″
Length over all—51′ 2″
Approximate Length Passenger Compartment—23′ 2″
Approximate Length Smoking Compartment—18′
Width over corner posts—8′ 6″
Width over all—9′
Height of car from top of rail over all—13′
Seating capacity—64
Seats—H. & K. Walkover type, 16″x34″, plush in Main Compartment and rattan in Smoking Compartment. Seats "A" are stationary.
Type of Trucks—Arch bar—Baldwin Loco. Works
Size of axles—6½″ motor, 5¾″ trailer
Wheel base—6′
Flange—M. C. B. standard
Tread—M. C. B. standard
Size of wheels, motor truck—36″
Size of wheels, trailer truck—33″
Wheels—Spoke wheels—Gibson shrunk on tires
Air Brake—Automatic, Partial AMM equipment
Hand Brake—Common staff type
Type and make of motors—West. 50 F.—550 V.
Motors—Two 150 H. P. each
Gear Ratio—19:52
Type of control—West. 251C U switch group
Lights—6 Circuits of 6 each
Current collectors—Trolley and third rail
Approximate weight of car complete—72,000 lbs.

Specification for Combination Passenger and Baggage Car is the same except as to interior arrangement.

Ford, Bacon & Davis,
115 BROADWAY
NEW YORK

This flyer was distributed by Ford, Bacon and Davis to advertise the surplus type L cars in 1918 and 1919. Despite extensive promotion, only three of the cars, including the combine, were sold, to the Galveston-Houston Electric Railway in 1920. *George Arents Research Library, Syracuse University*

This cluttered scene is significant in that it includes a rare view of work locomotive 301. The photographer was documenting dam construction work on Roaring Brook near the Spruce Street bridge in Scranton on October 18, 1917.
Edward S. Miller Collection

Changes were also made in the classification system at this time. Until then, large numerals designating the run numbers of regular trains had been placed on the front of cars or locomotives. The new system eliminated these numerals but required limited trains to carry a LIMITED sign by day or white classification lights by night. Dunmore division trains had to carry an EXTRA sign by day and yellow lights by night.

Murphy had not given up his desire for a derrick car for use in various maintenance-of-way projects and possibly to assist in clearing wrecks. In March 1919, he wrote to Zelnicker Supply Company of St. Louis, inquiring about three-drum hoist equipment complete with derrick with 500–600 volt motor to be placed on one of the two Rogers ballast cars which had been purchased in 1906. Murphy specified the crane should have a 30-foot boom, a maximum height of the A frame of 10½ feet above the car floor, and a ¾-yard clam shell, with the derrick having a lifting capability of three tons.

Former ash car 802, which was used for the equipment, was 9′6″ in width and had a length in front of the cab of 28′. Ford, Bacon and Davis located such a hoist equipped with a General Electric 40-hp motor from the firm of Hubbard, Floyd Company for $2,000. The only stipulation was the company insisted on full payment before delivery. Murphy was so pleased that he agreed readily to the terms, saying, "We will see that the bill is promptly paid and not held up due to the usual delay in passing through the office." The rebuilt car went into service by the end of the year and remained on the roster until 1954.

Early on the morning of January 13, 1920, combination car 113, which was standing in the yard at Scranton, caught fire, causing considerable damage. Combine 101, located on the adjoining track, was partially damaged by heat and water. It was necessary to summon the Scranton Fire Department to extinguish the blaze. The equipment was insured and Murphy wrote to the American Car and Foundry Company's Jackson and Sharp plant in Wilmington, Delaware, for an estimate for the repair of car 113.

The builder quoted a price of $6,800 which included providing a new interior mahogany finish and headlining, repairs to the woodwork above the windows, a new roof, seat cushions and backs, sash, and curtains; and refinishing the floor, hardware and basket racks. The entire car inside and outside would be painted and varnished in the usual manner. Jackson and Sharp would also rewire the car for lights and heaters, provided the L&WV supplied the fixtures. They offered to provide an entirely new carbody, which would be a duplicate of the type L combines constructed in 1902, for $10,540. The builder was willing to make a steel carbody but advised that it would be uneconomical to construct only one unit. Murphy decided to have 113 repaired as Jackson and Sharp proposed. The car was sent to Wilmington in February and returned in midsummer.

Indicative of the muddle in which the line's managers found themselves was the decision early in 1918 to sell some of the cars. At the time, the line had a total of 30 cars, 14 type L's and 16 type M's. Included in the total were four type L combines and three type M combines. Murphy had to check the records to determine which group of cars was the older. He suggested the line might dispose of one combine and seven coaches, all type L. Specifications were sent to various equipment dealers with an asking price of $10,000 to $11,000 per car.

Few lines were interested, probably because of the weight of the cars and their obsolete motors. The Hydro-Electric Power Commission of Ontario, for example, which was seeking three cars, rejected the Laurel Line's offer, even though Murphy cut the price to $9,500 each. Eventually, in mid-1920, the company succeeded in selling coaches 2 and 4 and combine 103 to the Galveston-Houston Electric Railway in Texas.

In June 1914, a growing coal trade forced the company to purchase 15 new steel hoppers from the Pressed Steel Car Company of McKees Rocks, Pennsylvania. They were numbered 751–765 and cost $13,800 in total. The next adjustment came in 1917 when Murphy decided that most of the 700-series cars were worn out and should be scrapped. Ultimately, however, six of these cars were rebuilt, using parts from the remainder and renumbered 711–716 and used as ash cars. One additional car which escaped the torch was 702, which was rebuilt as a work car 302, using the electrical equipment from one of the obsolete type L's.

Ever on the alert for a bargain, Murphy unwisely decided to purchase 20 old wooden gondolas from the Erie Railroad for $150 each. As part of the deal, the L&WV agreed to place its steel hoppers into revenue service and to minimize the use of Erie equipment between the Dunmore washery and the Quinn Coal Company breaker further down the line. These cars were assigned numbers 721–740, but by 1921 six of them were worn out and scrapped. The only other freight equipment was two Rogers ballast cars, numbered 801–802, purchased in 1906 for $700 each. As mentioned previously, 802 was converted to a derrick car in 1919.

A Rash of Accidents. Until now the safety record of the Laurel Line, considering its high speeds, wooden cars, frequent headways, intermixing of freight and passenger trains, and the absence of an automatic signaling system, had been impressive. This situation was to change for the worse, especially during the wartime years of critical labor shortages. The first serious accident occurred on December 29, 1914, when an inbound L&WV car struck a Wilkes-Barre Railway Company streetcar on the Hudson line at the North Main Street crossing shortly after noon.

Evidently, the L&WV watchmen neglected to lower the gates as the streetcar approached, thinking it had time to clear the crossing before the interurban arrived. Instead, the heavy Laurel Line train struck the side of the streetcar, derailing the rear truck and driving the car onto the sidewalk. Nine people were injured on the streetcar, only one seriously. As a result, the traction company ordered its employees in the future to ignore the L&WV watchmen and to rely on the conductor to flag the car across. Eventually, the L&WV paid $434 in settlement to the injured.

Less serious was a fire on car number 11 at the end of May 1916. This car was sitting on the station siding at the Wilkes-Barre terminal when a severe electric storm passed over the city. Suddenly, fire broke out in the smoking compartment at the point where the control cables entered the car leading to the switchboard. Station personnel led by Murphy quickly applied fire extinguishers to the blaze, which was finally brought under control by the city fire department. Murphy theorized that, since the trolley pole was down, lightning had struck the track and surged up into the car, causing the cable to burst into flames. It cost $1,200 to repair the damage.

No sooner had the excitement passed then a potentially more serious accident occurred on June 2. Late that night, locomotive 402 and three cars of coal coming from the Dunmore washery ran away on the steep hill approaching the Scranton terminal. The train hit the derail at a speed of about 30 miles per hour, jumped the track and came to a stop 160 feet distant. No damage was done and the locomotive and cars were rerailed by early morning. An examination revealed the brakes were in good working order and a test run with the same train indicated no problems. Therefore, the motorman was blamed for failing to keep his train under control and was suspended for 15 days. The accident did show the wisdom of installing the derail on the approach to the terminal loop.

A similar accident involving a passenger train took place two weeks later. In this instance, the 9:15 P.M. train from Wilkes-Barre via the hill route approached the derail at too great a rate of speed, with the result that the first car and the front truck of the second car left the track. There were no injuries or damages to the equipment but the motorman was suspended for 10 days for running at excessive speed.

More serious was a freight train wreck at Rocky Glen early in January 1917. A freight extra consisting of locomotive 402 and empty cars were running from Dunmore to the Gravity siding near Rocky Glen. This train stopped on the crossover just south of the park. The flagman had gone back about 500 feet when a regularly scheduled freight, consisting of express car 204 and one work car, running 20 minutes late, rounded the curve and struck the stopped freight head-on. The entire front of car 204 was shoved in, the rear truck driven to the front, and the front truck thrown out into a ditch. An NYS&W steel hopper was heavily damaged and the lower frame of locomotive 402 was broken across the king bolt. Both tracks were blocked for the rest of the night and were not cleared until 9:30 the following morning.

During this period, trains stopped on either side of the wreck and passengers had to walk around it. Three crewmen were injured, none seriously. Since the line lacked a crane and because the steam railroads were utilizing large cranes which would not clear the third rail, the wrecked cars had to be jacked up by hand and removed very slowly. Total damages amounted to about $3,000. Because of this and other accidents the company would eventually secure its own crane.

Another unlucky freight was a coal train at Dunmore in March. While moving slowly in the dark, an L&WV steel hopper split a switch and derailed under the DL&W bridge. Two old wooden cars that followed it were crushed by the steel hoppers behind. There were no injuries and the two wooden cars were immediately scrapped. Cost of other repairs amounted to $200.

The runaway jinx struck again in June. The same engineer was in charge who had been involved in the runaway the previous June. In this case, locomotive 402 and nine loaded coal cars came down the hill toward the Scranton terminal at an excessive rate of speed.

It crashed through the derail but damaged it so that all of the cars either rerailed themselves or remained on the track. Only one set of the wheels on the locomotive was on the ties. The track, however, was badly ripped up and was not repaired until 10:00 the next morning. The engineer was suspended for 45 days and removed from the motormen's roster. He was evidently using only the straight air brakes on the locomotive and hand brakes on the cars. This accident illustrated the difficulties the line was having with its inexperienced workers.

Locmotive 402 was beginning to get a reputation as being a jinx, since it was involved in another accident in August 1917. The train was loading brick at Hilldale and in order to protect it from the regularly scheduled freight northbound from Wilkes-Barre, a trainman was left at Plains substation. Unfortunately, this individual, who used to be an operator, went inside to assist the regular operator, who was receiving

These two views of the Dupont station were exposed January 5, 1917, before it was damaged by fire. Photos of this type were frequently taken for accident record purposes. One can only speculate that the engineer of a northbound limited or freight train rounded the curve onto the bridge over what later became the original US route 11, and found he had insufficient distance to stop before hitting someone on the track. Scranton Railway's bridge over the Lehigh Valley and Delaware & Hudson railroads appears at the far right of the view looking north.
Edward S. Miller Collection

an order from the dispatcher. While thus engaged, the local freight, headed by car 201, appeared. The trainman carelessly called out the window for the motorman to ''look out for the

railed, resulting in minor damage. The crew claimed that brake shoes on some of the hoppers were broken off, but an

an explanation from the crew. The only answer was for him to "go to hell." One can imagine his response! The power plant foreman complained that the practice of switching cars without ascertaining whether the track was clear had become routine. Again this reflected inadequate management control and the poor quality of help available during the war.

Not only were workers endangered by the cavalier attitude of crewmen, but property was damaged as well. In January 1918, an exasperated assistant engineer reported that the 7:50 A.M. train from Wilkes-Barre left the terminal without its conductor. The motorman had reached River Street before he discovered the absence of this vital crew member. He then backed up almost to the depot without bothering to change the trolley poles, thus tearing off four trolley hangers from the overhead. According to the engineer, ". . . such incidents are looked upon as highly humorous by the trainmen, at least judging from the laughter with which they greet such an event." He concluded this report with the plaintive statement, "If proper cooperation cannot be had from the transportation department in breaking up such practices, the maintenance of the line is a hopeless task."

A situation caused by carelessness of another railroad's employees was a fire on the viaduct over the Lehigh Valley Coal Company's tracks near Midvale. When the 1:50 P.M. train from Wilkes-Barre reached this point on May 10, 1918, the trestle was in flames, evidently caused by errant sparks from the coal company's locomotives. One of these engines had a spark arrester but it was not in position and the second one lacked the device entirely. Due to the high winds, about 100 feet of ties were burned and traffic on the L&WV was delayed for two hours.

To add insult to injury, the company received a bill from the Wilkes-Barre Fire Department for $74 for its services in extinguishing the blaze. Murphy estimated the total damages at $1,500. He urged the Lehigh Valley Coal Company to permit the line to tap a nearby water pipe so that "If the structure was set on fire in the future it would not be necessary to call out the fire department."

Thus far, the line had been lucky. Despite its poor safety record, there had been no fatalities during the war years. This situation changed dramatically on June 14, 1918, when a coal train crashed into the rear of a work train about 100 yards from the Moosic station, killing two and injuring 11. The work train was pulling out of the gravel pit siding about 4:30 P.M. when northbound accident-prone locomotive 402 with two cars of coal exploded two warning torpedoes placed 675 feet from the switch.

This was only three-quarters of the distance required under company rules, and since the track was on a downgrade the

adjacent Erie track. Damage to the equipment was heavy. Work car 301 was completely wrecked, with only the trucks remaining; the body was later loaded on a flatcar by the derrick and taken to the shops for rebuilding. The front end of unlucky locomotive 402 was demolished, but the trucks and underframe were undamaged. The flatcar and two steel hoppers also sustained minor damage. Murphy estimated the total cost of the accident to be about $20,000. The negligent flagman was discharged for violating rules and short flagging.

These accidents and many other minor ones were a financial burden as well as damaging to the line's reputation. Until 1918, it had been policy to set aside about two percent from transportation revenues for accident costs. In 1918, Murphy noted that about $15,000 had been accumulated, but over $30,000 had been spent as a result of accident claims and damages. Therefore, he suggested that four percent of the revenue be set aside for this purpose. Even this would be insufficient as the next decade would be ushered in by the line's worst wreck.

Finances Improve. In contrast to the early years, the second decade of the century showed an improvement in the financial condition of the Laurel Line. Management remained split geographically, with the officers scattered among several cities. Sproul, the president, played a nominal role. He remained an active figure in Republican Party politics, operating from his stronghold in the city of Chester, Pennsylvania. From 1919 to 1923, he served as governor of Pennsylvania. The treasurer, H.C. Kochersperger, maintained his office in Philadelphia. Residing in New York City were Secretary S.C. Stivers, and Chairman of the Executive Committee Frank Ford. On the scene in Scranton, actively directing the affairs of the line, were Vice President and General Manager Patrick Murphy and Superintendent J.H. Murray. Among the directors were, in addition to the officers, James Torrey, T.F. Penman, M.E. Yost and W.E. McDonald.

Generally, as indicated in the chart nearby, the company earned a net profit during these years. Part of it was distributed to the preferred stockholders and, beginning in 1919, common shareholders received dividends for the first time. Thus the financial reorganization effected in 1913 was successful, at least temporarily. Total investment in road, equipment and physical property was about $7,300,000. Income ranged from $275,000 to $415,000 and the percent of return on investment from 2.5 to 5.5.

One situation which did not change was the persistent question of free passes. In keeping with common practices of the railroads of the day, the L&WV solicited passes for its officers from the various steam railroads and freely issued passes at the request of other lines. In 1916, Murphy became

ment. After due consideration, Ford decided to leave well enough alone and passes were granted as before.

Murphy revived the issue two years later, this time with the Erie Railroad. He received a request for five passes for shop employees of that line, but he refused to fill it. He told the federal manager of the Erie that the L&WV had already issued 16 card annuals, 42 book annuals and one book annual good for five men. The book annuals were used by those who rode frequently over the road. In return, the L&WV had received only 10 card annuals for its employees and these were little used. He complained the Erie had as many book annuals as the DL&W, Lehigh Valley, Jersey Central and D&H combined and he noted that over 80 percent of the business of the L&WV was derived from passengers. He therefore advised the Erie to purchase 500-mile books good for one year at two cents per mile.

A wide range of companies furnished passes to the L&WV management. Aside from the obvious ones, they included the Clyde Steamship Company, the Merchants and Miners Transportation Company, the Wilson Line, Adams Express, Lehigh and New England, and Philadelphia and Reading.

Another persistent practice, which carried over from the Westinghouse days, was the tendency of the line to be remiss in paying its bills. The company files are cluttered with letters disputing the prices of vendors, references to misplaced bills and requests for duplicates, and dunning letters from unfortunate suppliers, complaining of nonpayment. Several of these were for significant sums. For example, the M.H. Ditrich Company of Chicago demanded $200 which was overdue for a year. The American Safety Lamp and Mine Supply Company had to beg for $767 owed to it. The federal manager of the Lehigh Valley Railroad angrily insisted upon a prompt payment of $2,200 for freight, which was several months overdue. He said he had written to the auditor several times, but he had never received a reply.

Murphy, in a favorite ploy, said he had sent the check "yesterday" and blamed the problems on a lack of communication between the Wilkes-Barre freight agent and the auditor. Even Ford, Bacon and Davis had difficulty collecting its monthly management and engineering fees. Evidently, the office force of the line had slipped into a pattern of carelessly meeting the company's obligations.

On balance, the Lackwanna and Wyoming Valley survived the wartime decline in reasonably good shape. Murphy, despite some faults, helped to bring stability and a less flamboyant sense of direction to the firm. Financially, conditions had improved and the prospects for the future seemed reasonably bright. The passenger equipment was getting old and rundown and although expensive repairs had been made, replacements would soon be necessary. Although the automobile was to prove increasingly troublesome, ridership had increased and the line was now taking a more realistic view of its prospects as a freight carrier. The euphoric dream of Charles Conn of the L&WV as an important link in a bridge route to New England had been replaced by the sober and unexciting realization that the company was essentially a terminal switching operation. The next decade would see the Laurel Line reach its peak years, but it would also be marked by drama and tragedy as well.

CONSOLIDATED INCOME STATEMENT

Scranton & Wilkes Barre Traction Corp.
Lackawanna & Wyoming Valley Railroad Co.
Lackawanna & Wyoming Valley Power Co.
(Year Ending Dec. 31)

	1915	1916	1917	1918	1919	1920
REVENUE						
Passenger	542,039	605,700	573,945	631,697	808,719	923, 528
Freight	75,992	65,633	97,428	125,976	153,633	190,318
Other	46,062	69,065	69,832	75,746	130,712	172,735
Total	664,114	740,399	741,206	833,420	1,093,064	1,286,582
EXPENSES						
Maintenance, Way & Structure	50,126	68,737	62,765	63,507	99,648	118,391
Maintenance of Equipment	55,710	55,049	67,409	92,878	129,414	134,652
Power	71,814	94,037	83,274	96,562	134,820	190,799
Transportation	128,646	127,399	139,209	169,872	190,370	228,348
Other	45,624	75,045	80,670	89,676	98,928	179,189
Total	360,924	420,269	433,329	512,490	653,181	851,380
NET REVENUE	303,190	320,129	307,877	320,930	439,883	435,202
Taxes	11,000	13,135	17,764	21,422	27,724	46,369
GROSS INCOME	287,282	313,562	288,917	288,914	393,309	352,511
Interest	208,383	208,283	209,084	207,152	208,506	206,724
NET INCOME	78,898	105,279	79,832	81,763	184,803	104,787
DIVIDENDS PAID	42,000	42,000	45,696	48,804	139,087	145,216
SURPLUS	36,898	63,279	34,136	32,959	45,715	571

Source: Financial Report, L&WV Railroad, 1915-1924, L&WV Collection.

The apex of Laurel Line equipment was the fleet of steel cars ordered in the 1920's, exemplified by this coach on the Minooka branch under the control of Joseph McNamara. *Edward S. Miller*

4

Decade of Extremes: Notoriety and Decline in the Midst of Normalcy

IN NOVEMBER 1920, with women included for the first time, Americans went to the polls and elected Warren G. Harding as the 29th President of the United States. Swept into power with him were Senators and Representatives from the Republican Party, which was to dominate the national political scene until the decade had passed.

Tired of wartime sacrifices and Wilsonian idealism, the nation entered a period popularly characterized by such diversions as jazz, flappers, high living and bathtub booze. New industries, including radio, the movies and the automobile, grew rapidly and helped to foster an air of prosperity. Thousands were engaged in stock market speculation, buying and selling in a frenzy that would culminate in the great crash of the market in October 1929. Looking deeper beyond the surface prosperity, historians have detected many signs of impending economic difficulties. Wages were relatively low; unemployment was troublesome and the farmer was experiencing growing distress. Once the appeal of the glamor industries of the day, such as radio and motion pictures, wore off, the basic weaknesses of the economy became readily apparent.

As with so many Americans, the fortunes of the Laurel Line varied during this time. A postwar recession was followed by a rapid recovery, which saw the line reach its peak in terms of revenue and passengers carried in 1924. After that, as paved highways spread through the valley, attracting growing numbers of automobiles and trucks, the railroad entered a phase of steady decline, which was a prelude to the years of the Great Depression.

The South Pittston Disaster. A harbinger of the line's worst disaster was a minor derailment in the Scranton yards in April 1920. Workmen were repairing the tracks when train No. 11 prepared to leave. One of the trackmen warned the motorman not to move his cars over a joint until he made it safe by applying bolts. The motorman carelessly replied he could get over that section with no difficulty and ordered the trackman off. When the rear wheel of the forward truck struck the loose joint, it caught on the rail and was thrown from the track. While minor, this incident revealed the indifferent and impatient attitude of crewmen on the Laurel Line. Seemingly insignificant violations of safety rules created an air of complacency, which would be an important contributing factor in the more serious wreck later that year.

On June 17, 1920, Murphy analyzed the safety record of the line in a letter to A.H. Ford of Hamilton and Wade, Inc., and made an unfortunate decision. His analysis indicated that the road had been in operation for 17 years, with the most serious accident costing the company ". . . a trifle over $10,000." He concluded that ". . . we probably will never have an accident to exceed $25,000." In response to Ford's suggesting insurance coverage ". . . to the extent of $100,000 in excess of $25,000, costing approximately $950 per year," Murphy said he had set up a fund of $50,000 for protection against a serious accident. Any additional coverage would be a waste of money and afford "very little protection." Within a few short weeks, this decision would prove to be among the most unwise ever made by the management of the line.

The July 4 holiday in 1920 as usual drew large crowds to the parks along the Laurel Line, with resultant frequent service, particularly at the dinner hour, when many of the fun-seekers headed home. Early in the evening of Saturday, July 3, passenger train No. 6 left Wilkes-Barre northbound for Scranton. The 6:10 P.M. run consisted of two coaches under the control of motorman Moran and conductors Cawley and McGrail. It was a normal trip until the train reached Inkerman, where a large holiday crowd from Valley View Park came on board.

As the train left Inkerman at 6:28 P.M., running about five minutes late, conductor Cawley instructed his motorman to run at slow speed until he finished collecting fares. Under those orders, they lost eight minutes in slightly less than one mile. Coming into the cut approaching South Pittston, the motorman got the bell to proceed and began to accelerate. Meanwhile, passenger train No. 21, car No. 11, a limited consisting of one coach, had left Wilkes-Barre at 6:25 P.M. Motorman White and conductor Mulrooney brought it into Plains on time. It left that station at 6:32 P.M. and quickly accelerated to 35 miles per hour.

This speed was maintained as No. 21 entered the cut at South Pittston, a place of limited visibility. Motorman White had been given no prior warning of the slow running train 6 in front of him, and he suddenly came upon it just as it was starting to pick up speed to about 20 miles per hour. White quickly cut the power and applied the brakes, but his single-car train struck the rear coach of No. 6 with a wrenching blow.

Some damage was done in this collision. It broke the air line and set the brakes on the 6:10 P.M. train and damaged the rear of its second coach. The vestibule of the limited was also damaged. None of the cars was derailed. Two individuals, both company personnel, were the only ones injured. White, on the limited, received a severe cut across the bridge of the nose, which bled profusely. Conductor Cawley, on No. 6, who violated the rules of the road when he left the first car and was standing on the rear steps of the second car as the collision occurred, was knocked off the train. He received a fractured skull, which was to prove fatal.

About the same time, northbound train No. 3, consisting of two coaches and operated by motorman Dailey and conductors F. Ratchford and Boone, was scheduled to leave Wilkes-Barre at 6:32 P.M., by then running four minutes behind schedule. At the Yatesville crossover, about 1,600 feet south of the accident, it was flagged by conductor Mulrooney of the limited, who boarded the train and rode to the scene of the accident. No. 3 then picked up conductor Cawley, who would die on the way to the hospital, backed to the crossover at Yatesville, took on board its own flagman, and proceeded north to Pittston, running on the southbound track. The dispatcher had given verbal authority for this movement against the current of traffic.

Up to this point, the situation, while regrettable, had been properly handled. The conductor of No. 21 had given flag protection, the injured man was being taken to the hospital as speedily as possible, and the dispatcher knew there was trouble on the road. Now, however, a series of errors of omission was made, which soon turned a serious but relatively minor accident into a major disaster.

While train No. 3 was performing its mission of mercy, the crew of No. 6, having determined that the first coach of their train had received no damage, decided to uncouple it, run it to Pittston with a load of passengers and then return for the passengers of the second car. They took the first step in this process by separating their two coaches and moving the first ahead a few feet. This work was interrupted by the sudden arrival of train No. 4, which touched off the calamity.

The two coaches of this train, manned by motorman Schlager and conductors J. Ratchford and Gabriel, had left Wilkes-Barre at 6:50 P.M. with no knowledge of the stalled trains in South Pittston. It reached Plains and Inkerman on time. For unknown reasons, the crews of Nos. 6, 21 and 3 had neglected to post another flagman or even to place torpedoes on the rail. Therefore, as the train entered the cut and descended the two percent grade, Schalger had a view of only 400 feet. At 7:05 P.M., the horrified motorman of No. 4 saw the three standing cars directly in front of him. Instinctively, he threw the brakes into emergency and leaped from the cab, as his heavy train, traveling an estimated 40 to 45 miles per hour, crashed into the rear car, train 21, which was filled with unsuspecting passengers.

The first coach of No. 4 rode over the underframe of car 11, splintering its wooden sides, shattering the glass, and telescoping it to within five feet of its front. The passengers and seats were compressed into a bloody tangle, as the sides of the mangled coach spread outward. The force of the collision carried both trains about 200 feet north of the point of impact, until they came to a halt lodged against a bridge. The lead coach of No. 4 was severely damaged, with all the seats

Coach 20, seen on a siding at the Scranton shops in July, 1920, was one of the type M cars involved in the disastrous wreck near Pittston that same month. *Harold Machler print, Edward S. Miller Collection*

The three cars most heavily damaged in the South Pittston wreck are illustrated on this page. The principal combatants (ABOVE) were the very heavily damaged car 11 and car 12 which overrode number 11. Note that the roof of number 11 remains almost completely inside number 12. The left elevation of car 11 (RIGHT) shows the extent of the damages which prevented repair. The steel center sill was designed to withstand normal buff loads but was distorted by the weight of the second car riding over it. Car 6 (LOWER RIGHT) had been moved inside the shop for repairs. The damage estimate for the five cars wrecked in this July, 1920, accident came to $12,760, which would approach one million in contemporary dollars. Not insignificant was the fact that almost one-fifth of the passenger car fleet was suddenly sidelined.

ALL: Edward S. Miller Collection from Charles Sobeck and Harold Machler

in the front section torn from their moorings and debris from coach 11 partially filling it. Sixteen passengers were killed and 20 passengers and two crewmen were injured, some seriously. Investigators later concluded that all of the fatalities and most of the injured were in the ill-fated coach of No. 21.

The explosive sound of the impact startled the local populace and news of the disaster quickly spread. Fire companies responded to the alarm, along with policemen and volunteers. Using axes and crowbars, they carefully extracted the dead and injured. Both private cars and ambulances took the survivors to Pittston Hospital and the dead to the local morgues. Many of the latter were difficult to identify because their clothing was shredded and the bodies were mangled. Miss Esther J. Tinsley, superintendent of Pittston Hospital and a passenger aboard one of the wrecked trains, escaped with a few cuts and bruises, while many in the same car were killed. She supervised care of the injured at the hospital throughout the night and almost 39 years later, on January 22, 1959, played a similar role in the Port Griffith mine flooding disaster.

Responsibility for this terrible collision involving trains 21 and 4 can be placed upon several employees. Conductor Mulrooney, on train No. 21, after notifying the dispatcher of the first accident, flagged No. 3, but he did not take any precautions to protect his train by flag after No. 3 moved north on the southbound track. Perhaps the excitement of the first accident confused him. Motorman White on train No. 21 knew that Mulrooney had returned to his train and that No. 3 had switched to the southbound track, but he did not take adequate measures to protect his own train either.

When No. 3 crossed over, flagman Boone of that train failed to put down torpedoes, nor did he consult the crew of No. 21 about the need for further protection. Thus, the crew of No. 3 contributed to the disaster by not taking any action to warn later trains of the presence of stalled No. 21. Dispatcher O'Malley also must bear some of the blame. He failed to take effective measures for the safe operation of the following trains, although he had been informed of the first collision. In the aftermath, Vice President Murphy believed any of these employees, by exercising good judgment, could have prevented the second collision. He might have added that his own decision not to acquire the $100,000 in liability insurance a few weeks earlier was also a serious mistake and it would make the long period of litigation involving the accident that much more trying for the Laurel Line.

Perhaps the most important cause of the tragedy was the absence of a block signal system. Considering the high speeds and frequent headways with which it operated, the Laurel Line had been extremely fortunate in the past. The technology had long been available to provide this basic safety feature and the management must be considered negligent for failing to install it.

Appalled by the tragedy, Murphy signed a contract with the Union Switch and Signal Company for an automatic block signal system along the main line, consisting of 14 color-light signals along the southbound track and 17 signals on the northbound track. The section between Connell Junction and Avoca was activated on August 15, 1921, and the remainder was placed in service on December 5, 1921. For several years, the staff system in the tunnel remained in use, until it was replaced with four additional block signals.

On July 12, M.J. Flatley, the master mechanic, provided Murphy with an estimate of damage to the five cars involved in the wreck. Four units could be repaired: car No. 5 for $125; car No. 6 for $375; No. 8 for $160; and No. 12 for $1,000. Car No. 11 was totally destroyed in the collision with a loss of $11,500.

By July 27, Murphy had information pertaining to the injured and to the dependents of the dead. He believed the company would be able to settle these claims for approximately $100,000. By that date, three lawsuits had actually been filed against the line with three or four others pending in the hands of lawyers. Murphy felt ". . . at least 90 percent" of all the cases could be settled out of court. His estimates of the amount involved in settling claims and the percentage of cases settled out of court were accurate. In a report to Frank R. Ford, chairman of the executive committee, he said the cost would slightly exceed $100,000. Most cases had been, or would be, settled in this fashion.

For some, however, the tragedy lingered for an extended period of time. Perhaps the most pathetic case was that of Mrs. G.J. Griswold. She had suffered a broken back, had a serious infection, was totally paralyzed below the waist, and the doctor's prognosis indicated she could die at any time or she could have lived for a number of years. Mrs. Griswold was still alive in the spring of 1921. Indications were that she could possibly survive for 15 years. Her family demanded a fixed annual payment to care for her throughout her lifetime. A settlement finally was made with an $11,750 cash payment and a $3,250 annuity.

Murphy apparently learned a great deal from the accident. He sent a letter to all trainmen employed by the Laurel Line saying that the accident ". . . should be a warning to every trainman in the service for the rest of his natural life." He believed it could have happened to any group of employees working for the road, since everybody had been ". . . depending upon someone else to protect them." For this reason, he was lenient with the men involved, ". . . realizing that they have suffered, knowing that they have contributed to such loss of life and injury." Murphy understood his own negligence in not providing adequate insurance coverage. He not only acquired a public liability policy for $100,000 but also increased coverage on all buildings and equipment from 80 percent to 100 percent. Fortunately, another major accident requiring this high protection would never again occur on the road. Unfortunately for Murphy and the L&WV, the brief period without coverage had exacted a heavy human and financial cost.

Other Accidents Mar the Record. A number of other accidents of a less severe nature occurred during this period. Shortly after midnight on November 22, 1921, the late train from Wilkes-Barre ran into a mine cave-in at No. 14 station, located between Hilldale and Inkerman. The train, consisting of cars 112 and 21, derailed, injuring several passengers. The Pennsylvania Coal Company had been working for several weeks removing pillars of coal in the vicinity of the right-of-way. They had informed the L&WV of this operation and supposedly an agreement had been reached whereby the miners would not approach within 35 feet of the track without notifying the Laurel Line.

Somehow, a misunderstanding developed and pillars were removed, causing a collapse six feet deep and 30 feet wide

The installation of automatic spacing signals, such as the one shown in this 1945 illustration, permitted safe operation of Laurel Line trains on very close intervals. Previously, the railroad's only signals were lower-quadrant semiphores at interlockings, presumably at Laurel Junction, Scranton Junction and Connell Junction. Train order signals, or order boards, were mounted at agency stations in Plains, Pittston and Avoca. The locomotive shown in this view below Connell Junction was involved in a collision with car 112 near the Scranton Tunnel in 1922, in a zone where signals had not been installed. *Edward S. Miller*

under both tracks. The only reason the accident was not more serious was the placement of a slow order when work began. As a result of this mishap, new procedures were adopted under which the L&WV engineer not only worked closely with his counterpart at the Pennsylvania Coal Company, but he also examined the mine workings himself. The cost of repairs to the cars and roadbed was nearly $2,800 and injury claims amounted to $2,700.

An accident similar to the June 1918 collision between a coal train and a work train near Moosic occurred in May of 1922, with much less serious consequences. A work train headed by locomotive 401 was moving slowly near the north portal of the Scranton tunnel, picking up old ties. A train from Wilkes-Barre, headed by car 112, ignored the flagman's warning and collided at moderate speed with the locomotive. Fifteen people were injured, none seriously. Motorman Frank Banks suffered severe lacerations and bruises. He later explained that, although he had seen the flagman, he did not anticipate the work train would be on the main line so soon after a limited had passed. Damage to equipment amounted to $1,500 and the estimate of the total cost of the accident was placed at $5,000. Once again, carelessness by employees was a major contributing factor.

Sometimes passengers were responsible for mishaps. For example, in July of 1923, a resident of Sugar Notch, ". . . very much under the influence of liquor," staggered toward a train in Wilkes-Barre. When the conductor yelled "All aboard," the passenger sat down on the edge of the platform, throwing his feet under the rear truck of the second car. The train was quickly stopped but not before it had crushed one of his feet at the instep, with the result that his lower leg had to be amputated. The accident report noted that "He had been drunk for several days, had quarreled with his wife and was in a rather ugly mood."

The old problem of runaways down the Scranton hill recurred on the afternoon of September 11, 1925. Train No. 11, car 23, on the Dunmore branch run, started down the steep incline at about 5:40 P.M. The motorman later claimed he applied the brakes with no effect, and the conductor put on the hand brakes, but they failed to hold. At this point, the motorman said he reversed the motors, but the train struck the derail at a speed of about 30 miles an hour and slid approximately 120 feet before it stopped. Happily there were no injuries.

Master mechanic Shellabarger and chief engineer Esslinger immediately examined the car and could find no defects. After No. 23 was rerailed, it was run to the top of the hill and descended in normal fashion, making five stops en route. Therefore, the management concluded the motorman had been negligent by permitting his car to attain too high a rate of speed before attempting to brake. He was suspended for 30 days.

An accident of a different nature hit the line late in the evening of May 19, 1921. The night operator reporting for duty at the Avoca substation discovered that the nearby freight house was on fire. Before the local fire department could reach the scene, two barrels of alcohol stored inside exploded and the entire building went up in flames. Insurance covered the $2,500 value of the structure and the $4,000 of freight and equipment stored therein. Murphy was uncertain whether the line should rebuild the freight station since the ground under the structure was being robbed of coal and a durable brick building could be affected by a subsidence. For the time being, a box car was stationed there and a temporary platform constructed. Eventually, a new building was erected at a cost of $2,200.

In August 1921, a tragedy occurred in the powerhouse. The top of the angle valve on boiler number one, which was being returned to service after repairs, cleaning and inspection, suddenly blew off. All the steam in the adjoining boilers rushed through the opening, scalding two repairmen to death. No exact cause could be determined, although the valves were analyzed by a DL&W chemist to see if the metal had become defective. Both men left widows with large, young families, who received a total of $3,900 in insurance and compensation. These and similar mishaps, mostly of a minor nature, damaged the line's reputation at a time when competition in the form of rubber-tired vehicles was increasing. To a certain extent, they contributed to the decline which began in the latter half of the decade.

The Great Train Robbery. The Laurel Line received unwanted national publicity as the result of a daring robbery on board one of its cars on July 30, 1923. All appeared normal in the Scranton terminal that morning as the 9:19 limited prepared to leave for Wilkes-Barre. Among the passengers were Arch Henshall, a paymaster for the West End Coal Company; two guards who accompanied him; and the usual mix of patrons, who were traveling south. Henshall was carrying the company's payroll of $70,126 in two metal containers, which he placed at his feet. He and the guards planned to change in Wilkes-Barre and board the 10:25 Pennsylvania Railroad train for their destination at Mocanaqua. Also boarding the car were five "swarthy looking" men, "apparently foreign speaking." One of them sat near the front of the car, opposite the coal company party.

As the train approached an isolated cut, south of the Moosic station, the man seated at the front suddenly jumped to his feet, opened the door to the motorman's compartment and fired a single shot through the chest of P.J. Durkin, who managed to bring the car to a halt before he slumped over. At the same time, the other four bandits, wielding guns, ordered the startled passengers to raise their hands. One young man, Edward Murphy of Scranton, a salesman for an oil company,

The newly erected replacement freight station at Avoca is the focal point of this photograph taken in 1922. The adjacent building is the fifteen year old substation. *Edward S. Miller Collection*

The taxi stand at the Scranton station was later enclosed to provide space for a soda fountain with, as the sign in the window proclaimed, tables for ladies. An armed forces recruiting poster swings in a steel frame beside the curb in this 1920's-era photo. *Edward S. Miller Collection*

The view from the corporate office in Scranton reveals the formidable wall separating the Lackawanna passenger station from the L&WV's facility. Negotiations with the DL&W for direct pedestrian access never reached fruition, as always because the Laurel Line was looking for a free lunch.
Edward S. Miller Collection

Glum-faced miners solemnly await a train at No. 14 station, Saylor Street at North Main Street, Plains, on July 21, 1929. The officious gentleman with the cane appears to be pointing to the poorly maintained cinder platform, which was typical of many L&WV way stations. He may have been indicating the point where a claimant stumbled. *William F. Ellis Collection*

was slow to comply and was fatally shot through the abdomen. A wild scene ensued. The two guards made a brief attempt to defend the payroll, were shot at but narrowly escaped injury. Paymaster Henshall received a wound in the thigh as did a young man nearby. A woman jumped off the rear platform and suffered lacerations and bruises. The conductor attempted to wrestle the gun from one of the bandits but another placed a pistol at his head, forcing him to back down. In all, one man was killed and three wounded in the melee.

The thieves seized the two money cases, jumped from the train and scrambled up an embankment to a waiting Buick touring car, next to which stood a short, stockily built man, also armed with a gun. The robbers fled the scene in the direction of Moosic. The conductor ran the car to Avoca; he phoned the police and continued on to Pittston where ambulances removed the dead and injured. A day later the getaway car was found abandoned on a dirt road in the nearby town of Taylor.

State and local police launched an intensive investigation but had no success initially. Eight months later, in March 1924, a man named Tony Burchianti was arrested near Homestead, Pennsylvania, during an unsuccessful $400,000 payroll robbery. An investigation revealed that he was connected with the L&WV crime and, in fact, had been employed by the West End Coal Company along with Toni Torti until they quit their jobs just 10 days prior to the robbery.

Torti's capture came in spectacular fashion. He and "Big Jack" Stummi, a confederate from the Laurel Line robbery, entered the post office at Wiltonville, Ohio, where state police, acting on a tip, were awaiting them. When the officers, led by state police Lieutenant Jacob C. Mauk, confronted the pair, they bolted for the door with Torti shouting, "We'll never be taken alive." Jumping into a car, they sped away, followed by the state troopers. A running gun battle followed. Eventually the suspects' car stalled, forcing them to flee through a field and take cover behind a shed. The police quickly closed in, wounding both men. "Big Jack" evidently committed suicide, but Torti was captured.

Returned to Pennsylvania, he and Burchianti were identified as the Laurel Line gunmen, charged with murder, robbery and assault and tried in Scranton in October 1924. The jury found them guilty of first degree murder and Judge George W. Maxey sentenced them to death. Several defense witnesses who had attempted to substantiate the alibis of Burchianti and Torti were charged with perjury and subsequently tried. One was sentenced to 3½ to 7 years in prison and two others were given suspended sentences. On June 1, 1925, Tony Burchianti and Tony Torti were executed in the electric chair at Rockview State Penitentiary in Bellefonte, Pennsylvania.

One other bandit, Paul Farci, was later convicted and executed for the Laurel Line holdup. A fifth member of the gang was involved in a murder in Cambria County and also executed. Thus, five of the six perpetrators paid their debt to society with their lives.

Operating Problems Challenge Management. Unlike the excitement of the robbery, most of the operating problems of the decade were mundane. In some ways, the Laurel Line continued its inefficient practices. Although an inspection report by the Mutual Fire, Marine and Inland Insurance Company of Philadelphia indicated that the property was in generally good condition, a number of recommendations were made, more than a third of them for the second time. The report noted there was a "general carelessness" about refilling and recharging fire extinguishers; fuses were exposed and uncovered; wood instead of metal was used in lockers and waste boxes; and there were no fire drills.

Some attempt was made to tighten inspection and operating procedures. For example, master mechanic Flatley wrote a detailed memo listing specific items that were salvageable and were to be reused by the shop forces instead of simply being discarded. He stated that ". . . the cost of maintenance is enormously high at present." Still, difficulties persisted. Murphy circulated a memorandum in 1921 calling the attention of employees to numerous claims presented by passengers for damage to clothing caused by contact with grease, tacks or other hazards in the coaches. The vice president and general manager asserted that the problem "indicated laxity on the part of the inspectors" and demanded immediate improvement.

When the insurance renewal on the carhouse and shops was made in 1921, Murphy felt $56,000 would cover the stock-

An L&WV section gang ground their implements while the photographer is in the area. The track under repair is the Dunmore line along Roaring Brook, east of the Harrison Avenue bridge in Scranton. Construction of the bridge dates this picture to 1921 and confirms that double track was still in use over the hill in the third decade of the century. *William F. Ellis Collection*

room, machine shop, paint shop and offices. In response to a question from the insurance agent, he reported the three locomotives and five cars carried Pyrene fire extinguishers, while the remaining cars had three-gallon soda acid tanks. He also indicated the location of cars at night between 1:00 and 5:00 A.M. Five cars were kept in the carhouse, four in Wilkes-Barre yard, 15 in Scranton yard and two in service. In the winter, five cars were shifted from Scranton yard to the carhouse.

There were the usual cases of near misses at grade crossings, some of them rather humorous. One motorist was cited by the company for passing the gateman at the North Street Crossing in Wilkes-Barre and darting in front of the 11:19 train, which, fortunately, was able to stop. Murphy wrote a letter to the offending motorist advising him to exercise greater care. The individual claimed innocence, indicating the watchman must have noted the wrong license number. Murphy answered by saying, "If your car number is 59458, then your car is the one that stopped on the crossing, compelling the 11:19 train to come to a stop. I, personally, was standing on the front end of the car and took the number borrowing a pencil from the motorman to do so." No more was heard from the now subdued motorist.

Providing taxi service at the Scranton terminal also challenged the ingenuity of Patrick Murphy. The earlier arrangement with the Scranton Taxicab Company had expired with the result that vehicles were available only on a sporadic basis. A local citizen offered to provide a taxi service and the line rented him a small office for $10 a month. Unfortunately, he was prone to drink; Murphy wrote that ". . . at times he absorbs about all the liquor in Scranton," despite Prohibition.

As a result, Murphy canceled the agreement with him and encouraged a group of employees to organize a firm under the name, "Laurel Line Taxicab Company," and even built a new office for their use. The problem was he did not consult the state Public Service Commission about any of these changes and when that agency learned of them it preemptorially ordered the line's officers to appear at a hearing. The matter was resolved, but again it could have been avoided.

Freight provided its share of operating problems. At the beginning of the decade, the L&WV asked its connecting railroads, the Erie, DL&W, and LV, to take over inspections of cars delivered to or sent by the electric line. As a reason, Murphy said this would permit the company to fire its own inspectors, ". . . who wanted a raise anyway and were likely to cause trouble." After extensive correspondence, the railroads agreed, although the L&WV was billed for their services.

Sometimers the high-handed manner of the L&WV got it into trouble. The company had ordered a carload of ties through George Bingham of Dunmore from the Valley Tie and Timber Company of Roanoke, Virginia. When the car arrived, the L&WV rejected the ties as inferior and eventually decided to sell them to Spruks Brothers, a local lumber company, for 50¢ each. However, Bingham never paid the Valley Tie Company and the L&WV was soon being dunned with demands for compensation. Since the Laurel Line would have to pay freight and unloading charges, its net loss was about $100. The situation could have been avoided by calmer management procedures.

Conversely, when the Wilkes-Barre Iron and Wire Works installed a foot bridge over the Laurel Line tracks at the Moffit Street School in Plains, it asked the L&WV to provide a car and derrick to assist in the operation. The job was completed with no difficulty, but the L&WV submitted an outrageous bill for the services of 15 men, even though the bridge company had its own crew. Wilkes-Barre Iron Works rejected this bill as excessive and pointed out that no more than seven men were seen on the L&WV train. Again, some common sense would have avoided this embarrassment.

In general, the 1920s saw a downward trend develop in both freight and passenger traffic for the L&WV. Increased competition from vehicles on new paved highways then being constructed in the area, strikes in the coal fields and changing public tastes regarding amusement parks were among the factors responsible. In 1920, the line carried 4,220,596 passengers for a revenue of $923,558. Ten years later, the road carried only 2,108,794 people for a revenue of $532,619. The decade, therefore, saw a 50 percent decrease in passengers carried and a 42 percent decline in fares collected. Fares were raised only twice in this period: there was a 20 percent increase in August 1920 for one-way fares and a 13 percent increase in March 1928.

Despite this serious fall in patronage, service was modified only slightly. In the World War I period the line was running

The City of Scranton had reached its peak population of about 144,000 in 1930 when Fairchild Aerial Surveys recorded this photo of the city from the east. The wooded area west of the railroads in the lower right corner is Nay Aug Park. The extensive L&WV land holdings had been acquired from Lackawanna Iron and Steel Co. Interstate highway 81 (from left to right) and the Central Scranton Expressway today obliterate most traces of the L&WV rights-of-way.

Edward S. Miller Collection

47 local and 12 limited trains in each direction. Locals stopped at all stations and averaged about 40 minutes for the one-way trip, while the limiteds, stopping only at Pittston and River Street in Wilkes-Barre, made the trip in 30 minutes. The limiteds ran every hour from early morning through early evening. Rush-hour headways were 20 minutes and base service headways were half-hourly.

This pattern was maintained until September 1929 when the number of limited trains was cut from 12 to five, three in the morning rush period and two during the evening peak. By October 1930, only two limiteds remained on the time card and these were gone by the following year. All trains, except the Dunmore branch locals, used the Scranton tunnel. The latter cars serviced the old main line over-the-hill route from Laurel Junction to Maple Street, as well as the Dunmore branch itself. The stretch of track from Maple Street to the jointure with the tunnel route at Connell Junction was used only by some freights and special movements.

Instability in the coal fields in terms of both labor and market conditions adversely affected the Laurel Line. When workers went on strike or were laid off, passenger earnings fell as much as 25 percent. During these periods, Murphy was unwilling to reduce schedules, except for running occasional single cars instead of two-car trains. Economies were made in the track department and shops by putting the employees on a five-day week instead of the normal six. Therefore, the line was able to effect only minimal savings despite precipitous declines in patronage.

Some thought was given during this period to substituting a gasoline motor car or "doodlebug" for electric cars on the Dunmore branch. A comparison study was made between Chicago, Rock Island and Pacific motor car number 2550, operating on a branch of that company, and the interurban cars of the L&WV on the Dunmore branch, for a six-month period. The maximum grade on the Rock Island branch was 1.25 percent versus 5 percent on the L&WV. Schedule speeds were comparable, 23 mph on the Rock Island and 24.8 mph on the Laurel Line.

Cost of operation was considerably lower for the Laurel Line, 6.4¢ per mile versus 9¢ on the Rock Island Line. Management concluded, considering the much greater number of passengers carried on the Dunmore branch, that conversion to gasoline motor cars would not be economical. It might be added that the motor cars would have had a far more difficult time negotiating the steep over-the-hill grade to Dunmore than the electric cars.

As usual, the L&WV had trouble deciding what kind of line it really was. In many instances earlier, the company had stressed that it was in every way equal to the steam railroads and just happened to be operated by electricity. However, in the early 1920s, Congress passed a transportation act, Section 15A of which dealt with excess railway operating income. The L&WV now did a flip-flop and argued before an ICC hearing that it was not a freight carrier but simply an interurban electric road and therefore exempt under the terms of the Transportation Act of 1920.

Murphy's real reason for preferring to be classified as an interurban was pragmatic. He pointed out in a letter to the American Shortline Railroad Association that the L&WV paid wages about 20 percent lower than prevailing steam railroad rates. He feared that if the L&WV were put in the steam railroad category, the shop, train service and freight workers would immediately demand higher wages. "This would be a very severe blow and would make it rather difficult for us to earn even five percent on our investment. We very much prefer to be classed as an interurban road. . . ." The ICC examiner felt differently and found the L&WV was engaged in the general transportation of freight. The company appealed to the full commission which decided against the carrier by one vote in January 1926.

Service to the Parks. Despite this unfavorable ruling on the line's status as a freight carrier, the L&WV remained primarily a passenger hauling line. In typical interurban fashion, an amusement park along the route generated considerable revenue, but as with everything it touched, the corporate involvement of the Laurel Line with Rocky Glen Park was perplexingly complex.

After lying dormant during the World War I period, Rocky Glen was reopened in 1919 under the auspices of the Glen Amusement Company, headed by A.J. Duffy. Glen Amusement, in turn, made an arrangement with the partnership of John Nallin and Joseph Jennings who had purchased the other half of the divided park for joint operation of the facility.

Many improvements were made in the park during this period. In addition to the old attractions such as the Swiss cottage, the merry-go-round and various stands, new amusements were constructed, including the whip, ferris wheel, penny arcade, milk bottle, and a shooting gallery. A major attraction was a roller coaster constructed by the Philadelphia Toboggan Company in 1920.

Indicative of the financial condition of the Glen Amusement Company was the contract under which the toboggan company retained ownership of the ride, paying 25 percent of the gross revenues to the park company. When Nallin and Jennings decided to add a new device, known as the "Dodgem," which cost $9,000, Duffy was unable to raise his share. He turned to the L&WV for a loan of $4,500. Murphy agreed to a contract, whereby the railway invested that sum and received all the revenue due to the Glen Amusement Company from the Dodgem until the loan was repaid with interest. At this point then the L&WV was directly involved in the operations of the park.

Indirectly, the Laurel Line had been involved with the sale of admission tickets to the park. Every year the road made a contract with the Glen Amusement Company whereby it agreed to sell round-trip tickets to Rocky Glen, including an admission stub for 10 cents. One additional person was admitted free with every admission coupon presented.

Generally, the L&WV benefited from its service to Rocky Glen. Until 1920, it was the only way patrons could get to the park. In that year a road was completed to the Nallin and Jennings portion and access by automobile was possible. Patronage at the park varied considerably each season depending upon such factors as the weather, strikes, working conditions and the whim of the public.

In 1921, Murphy estimated the line spent more than $12,000 in running extra cars or trains to the park but that it received an additional revenue of about $70,000 from ticket sales, leaving a net profit of about $58,000. In 1922, in contrast, expenses totaled $40,000, but revenue was only $37,000, leaving a deficit of $3,000. Murphy attributed this to strikes in the coal mines and claimed he had to maintain regular schedules and could only economize by taking off a car or two. Three years later, at the request of the Glen Amusement Company, the L&WV constructed a side track at Rocky Glen station for the storage of extra cars.

The Laurel Line station at Rocky Glen was originally constructed in 1904 on leased land. At first, it was basically a shelter with a roof and sides four feet above the floor. In the winter, the sides were completely enclosed. When the park reopened in 1919, A.J. Duffy requested that the building be used as a dance hall. This was done for three years but in 1922 it was converted to a penny arcade. Despite this activity, the building was still available for railroad patrons. A fire destroyed the structure in the fall of 1928.

By the following spring, the railroad completed a new building at a cost of $11,600. It was 100 feet long, 40 feet wide and 21 feet 8 inches high, with concrete block foundation, steel superstructure, composition sides and roof and a floor suitable for dancing. Until 1935, it was used as a dance hall and intermittently as a roller-skating rink, while still serving as a station for L&WV passengers.

Evidently, the Glen Amusement Company was unable to make money operating its portion of the park. The Nallin and Jennings half, on the other hand, remained solvent and was not involved directly with the L&WV, partly because the automobile road provided access to their gate. At the end of the 1927 season, Glen Amusement failed to meet its lease payments to E.S. Williams for the land on which the park was located. Murphy and George W. Esslinger then took over the obligation of making the lease payments and made a contract with the Glen Amusement Company to operate the park in the 1928 season.

By this time, Glen Amusement had obligations of about $150,000, including $5,000 in cash advanced personally by Murphy. At the end of the 1928 season the amusement company defaulted once more and Murphy and Esslinger decided to take over operations themselves. For this purpose they char-

Photographic coverage of the Laurel Line in the 1920s, like that of many electric railways, is spotty at best. The postcard and publicity cameramen had moved to newer subjects. The railway enthusiasts had not yet organized. In 1926, the Laurel Line erected a wide array of no trespassing signs along the railroad, which were dutifully photographed. This view at Rocky Glen Park, in addition to the new placard, also includes the roller coaster that once crossed over the interurban tracks.
William F. Ellis Collection

tered the Glen Park Company in May 1929 with an authorized capital stock of $10,000. Esslinger was elected president, Murphy treasurer and J.A. Chisholm secretary. Offices were maintained in the Laurel Line station in Scranton.

Murphy and Esslinger ran the park through the Glen Park Company from 1929 through 1935. During this period their control was challenged by Duffy, president of the Glen Amusement Company, who claimed his lease was still valid. The matter was taken to court, but before it was resolved Murphy attempted to physically eject Duffy and his friends from the Swiss cottage, which they occupied. An altercation took place between Murphy's officer and the Duffy forces, with the latter retaining possession.

Eventually, in 1939, the court upheld the claim of Murphy and Esslinger. Duffy still remained dissatisfied and later filed a lawsuit challenging the control of the duo. By this time, however, patronage had fallen sharply, partly because of the Depression, and the Glen Park Company had failed at the end of the 1935 season. E.S. Williams then sold the land to Benjamin Sterling, Jr., who reopened the park in 1936. The lawsuit was eventually dropped since Murphy and Esslinger no longer had possession. All told, it was a costly experience for the vice president of the Laurel Line, since he lost a considerable amount of money from his ill-advised venture into the amusement park business.

Finally—A Mail Contract. After many years of agitation, the L&WV finally succeeded in getting a mail contract with the Post Office Department in 1921. However, it proved to be a mixed blessing and soon produced controversy.

On August 4, 1921, Murphy reported to Ford that the post office would begin mail service on August 15. Postal authorities requested three southbound and two northbound trips be designated to carry from one to 10 sealed pouches of mail. This could be handled on regular passenger cars or baggage cars, as the company desired. Estimated revenue was $237.07 per year.

Almost immediately, a dispute arose over the method of handling the mail. Post office officials agreed to take the pouches to the Laurel Line stations in the terminal cities. They were not required to physically place the mail in the cars, but, at Murphy's request, they agreed to have their transfer clerk take the pouches to the car door. Murphy, sensitive to potential demands for additional compensation from the trainmen if they had to perform any extra tasks such as handling the mail pouches, felt the transfer clerk should actually place the mail in the car. He told Ford that in this way the company would not be responsible for handling the mail and "no claim can be made by our employees for any extra compensation." The difference in interpretation of what was required soon caused trouble.

In October, Murphy received a letter from J.T. Campbell, chief clerk of the Railway Mail Service in New York, complaining about the procedures used in the Wilkes-Barre station. He said he had recently been in Wilkes-Barre and was informed by the transfer clerk that the L&WV stationmaster insisted that the postal clerk go inside the car and pile the mail there.

The clerk had pointed out he was not obliged to do so. Campbell asked Murphy to instruct his workers to receive the mail at the car door. Murphy replied that the clerk had been "impudent" and had thrown the pouches on the floor where passengers were in danger of tripping over them. He said L&WV trainmen could handle the mail, but this would involve extra expense to the government.

Murphy was on very weak ground in his argument. Postal regulations specified that mail had to be delivered to the station of the railway company. The chief clerk had already bent the rules by having the pouches taken to the car door. He would go no further, nor would he be responsible for any added expense. After a conference with him, Murphy was forced to agree. Evidently, the trainmen consented to perform the minimal duty of stacking the mail without extra pay.

Coal Becomes the Bulwark. During the decade, efforts continued to develop the freight traffic. Interchange agreements had been reached with most of the major steam railroads with the exception of the New York Central. Murphy wrote to a vice president of the Vanderbilt system in October 1921 describing difficulties and inconveniences caused by the absence of through rates. When he received no answer, he wrote two months later to R.D. Starbuck, assistant vice president of the NYC, referring to his earlier letter and again pressing for an agreement.

Murphy felt the problem lay in the fact that the L&WV was electrically powered and therefore regarded as a mere interurban railroad by the mighty Central. Although he had pleaded to the ICC only a year earlier that the L&WV should be regarded as an interurban line, in order to avoid being placed under the terms of the Transportation Act of 1920, he now assured the NYC assistant vice president that the Laurel Line operated under a steam charter, had double-track, standard steam railroad construction over private right-of-way and large terminals in both Scranton and Wilkes-Barre. He asserted, ''We are better equipped to handle freight in the city of Wilkes-Barre on teaming track and freight house than any other line, with the exception of the Pennsylvania Railroad and we are equal in location to them. . . . We own and interchange freight equipment.''

Hinting at possible political favors, he noted that Pennsylvania Governor William Sproul was also president of the Laurel Line and ''has acted very friendly to railroads in the state.'' The NYC was unmoved by Murphy's arguments and replied that the question of through rates with electric lines was being considered by the Central Freight Association and no action would be taken until a decision was reached.

Rates were adjusted throughout the decade, generally in line with steam railroad changes. Tariff Number Eight, which went into effect on July 17, 1918, increased rates 25 percent. This was the first time L&WV rates fell below those of competing steam railroads, since those carriers were granted even higher increases for distances under 20 miles. In September 1920, all rates were increased 40 percent.

This drastic revision was too severe, particularly when combined with growing motor truck competition. Therefore, in July 1922, tariffs were cut 10 percent, remaining at this level for the rest of the decade. In October 1929, a 50 percent reduction was made on less-than-carload freight (LCL) for all commodities classified as groceries. They included fish, flour, fruits, juices, grains, milk and vegetables. The cut was necessary because of truck competition.

As early as 1920, W.L. Cairns, the general freight and passenger agent at Scranton, had alerted employees to the

The sidings at the W.Y. Moffat Colliery at Virginia are filled with loaded hopper cars in this late 1920's view. This branch from the main line at Virginia was built in 1920 and removed in the late 1930's. Carloadings ranged from a low of 7 in 1923 to a high of 4,769 in 1937. In the latter year, Moffat accounted for 37 percent of the L&WV carload freight. *Edward S. Miller Collection*

The L&WV was one of the very few electric interurbans to directly service coal mines. This 1926 photo, taken to show no trespassing signs, includes a trio of hopper cars beneath a coal tipple along the northbound track at the Prospect Colliery in Plains Township. The wavy alignment of the third rail exemplified the difficulty of maintaining track over coal mines.
William F. Ellis Collection

necessity of providing one-day service. He pointed out that the Scranton and Wilkes-Barre Express Company, a truck line, was making five trips a day and attracting LCL freight normally carried by the Laurel Line. He urged agents to expedite shipments and report any problems. By 1930, L&WV local rates were about 28 percent lower than those of the steam railroads between Scranton and Wilkes-Barre. Despite this, trucks were winning a greater share of the LCL market.

Sometimes the L&WV was unable to meet requests for service even though it wished to do so. For example, in September 1921, the Philadelphia Toboggan Company requested the electric road to supply one carload of good egg or stove coal to be shipped to Philadelphia via the Lehigh Valley and Reading Railroads. Murphy was forced to decline the offer, since there were no through rates to the Quaker City for independent miners located along the Laurel Line's right-of-way. The local rate would make the freight tariff so high that the cost of the coal would be excessive. He advised the company to try elsewhere.

An analysis of freight service for the month of June 1923 showed the L&WV interchanged 757 cars, which remained an average of about three days on its tracks. The nearby local lines naturally predominated, but cars of such diverse and distant carriers as the Santa Fe, Central of Georgia, Canadian Pacific, Northern Pacific and Union Pacific were handled. The L&WV was assessed $2,300 in per diem charges. This was partially offset by $274 in demurrage and by $750 due the Laurel Line in per diem charges for the use of its cars by its connections.

As was to be expected, the bulk of carload freight handled by the L&WV was anthracite coal. In a report to the ICC on commodities carried for the year 1928, the line stated it had hauled 1,228 cars of hard coal for a revenue of $28,000; 556 cars of fresh meat, which earned $10,400; 215 cars of clay, gravel or stone for a revenue of $5,400; 301 cars of motor vehicles, earning $5,000; and 177 cars of petroleum and oil products, which also produced $5,000 in receipts. In addition, 680 carloads of coal were hauled locally on the Laurel Line for a revenue of $19,100. Among the largest of the local shippers was the Moffit Coal Company near Virginia. In 1928, it shipped 1,363 cars of prepared coal, although in the following year car loadings declined to 1,008, which was the least amount of coal Moffit had produced since going into production. It was clear that the fortunes of the L&WV as a freight carrier were tied closely to the prosperity of the mining industry.

Labor Peace Prevails. The trend of employment on the L&WV was downward in the 1920s. In 1921, the line provided jobs for 369 people. The ranks of the officers and supervisors were gradually being thinned. There were five general officers, 12 clerks, two supervisors and 105 maintenance-of-way workers, one shop foreman and 50 shop men, two powerhouse supervisors and 64 workers, and one transportation superintendent and 127 personnel. In four years' time, the total number of employees had fallen to 228 and in 1927 it was 214. By 1929, employees numbered 160. There were some cuts in the supervisory personnel and major reductions in the number of maintenance-of-way and equipment forces. This downward trend would continue through the Depression years.

In general, labor peace was preserved during this period. As Murphy had feared, the trainmen did interpret the government's position on the status of the Laurel Line under the Transportation Act of 1920 as justification for demanding higher wages. He was willing to negotiate, but he felt the men were already earning a more than adequate salary. He told the chairman of the grievance committee of Laurel Lodge Number 711, Brotherhood of Railroad Trainmen, that if they were not earning sufficient money, they could work longer hours. He said, "The work you are doing is much lighter and less laborious than many other occupations and I see no reason why men in light occupations should confine themselves to working eight hours."

He was also reluctant to make suggested rule changes, which would increase the company's operating expenses. Murphy instructed the trainmen, "Trying to obtain something for nothing is very poor business and the great trouble with the county today is that too many are practicing these methods and are not giving a fair return for the compensation they are receiving."

The best the workers were able to accomplish by 1922 was a raise of about 3½¢ an hour. Motormen and conductors now earned 63½¢ an hour and brakemen were paid 58¢ an hour. The reason for the ½¢ differential was the union's willingness to permit one conductor to work two cars on certain trains between Avoca and Scranton. An additional 3¢ per hour was negotiated in 1923.

The working rules of the line guaranteed trainmen a minimum of eight hours' pay. This could be earned by putting in the time or traveling 150 miles. Extra men who were called were guaranteed a minimum of four hours or 75 miles for each call. Standard rules of seniority prevailed and promotions could be made through the ranks from brakeman to motorman or conductor. Trainmen were required to wear uniforms and motormen to wear overalls and jumpers of a standard style. Two brakemen were to be assigned to all freight trains consisting of six or more cars.

There were a few fringe benefits as well. For example, the company insured its employees for sums ranging from $500 to $1,000, depending upon term of service. A stock purchase plan was provided for employees. When the power company workers withdrew from their union in 1923, they asked if they could purchase some of the Scranton and Wilkes-Barre Traction Company stock to use as a saving fund. Murphy felt this was a "wonderful idea" and agreed to buy the shares and resell them to the men on a monthly basis. Eventually, 70 men were participating in this plan. As time would prove, the purchase of stock in the Laurel Line was questionable.

Even Murphy was sufficiently optimistic about the line's prospects to buy shares himself. When Westinghouse severed its last link with the L&WV in December 1921, it sold its remaining shares in the Scranton and Wilkes-Barre Traction Company. Murphy soon purchased 500 of these shares, including 250 sold by Ford and President William Sproul, for $2,645. Actually, he received "Voting Trust Certificates," since most of the actual stock was held by the Guarantee Trust Company of New York as collateral for the bonds issued by the traction corporation. By 1928, when the company was reorganized, Murphy owned 1,600 shares of common and 225 shares of preferred, which he exchanged for new securities issued by the L&WV.

Murphy displayed his paternalistic attitude toward employees in a case involving the widow of one of the workers. She was the beneficiary of his insurance and the vice president had arranged for her to receive $60 a month until the principal was exhausted. Instead, the woman withdrew $1,325 in one year, provoking Murphy to send her a check for the balance of $410, accompanied by a letter chastising her for her profligate ways. He said he had expected her to . . .

> . . . spread the money over a period of about two and one-half years, to enable the younger children to grow up and be kept in school. Instead of you doing this, a perfect mania seemed to come over you to try and get the money in the least possible time and squander it. I am sending the enclosed check to you, we are, therefore, wiping our hands of the whole matter, as we find you have two grown sons who should assume some responsibility. . . .

Modifications Along the Line. Track was occasionally upgraded during the decade. The company ordered 150 tons of 90-pound ASCE rail from the Carnegie Steel Company in February of 1920. Additional orders were placed in later years. Certain industrial sidings were extended or opened during the period. In May 1920, the siding at Midvale was extended to 845 feet to handle coal from the Prospect breaker. A new siding was opened to the Wilson and Company plant in Wilkes-Barre in January 1922. The former Wyoming Valley Brick Company siding at Hilldale, consisting of 997 feet of track, was purchased by the L&WV for $550 in 1924.

In all, the company had 58,295 feet of yard tracks and sidings, including privately owned spurs, during the 1920s. A number of these served mines or breakers such as the Minooka breaker, the Quinn Coal Company in Avoca and Midvale, and the Dunmore breaker. Scranton yard contained more than 9,000 feet of track, Wilkes-Barre yard 7,200 feet, and Pittston yard 800 feet.

A dispute with the Erie Railroad developed late in 1922 when the L&WV proposed erecting 200 feet of trolley wire on the interchange track at Dunmore. This would make it easier for the electric locomotives to push cars into the Erie yard tracks and would speed operations. The argument developed over the height at which the trolley wire was to be strung. Standard clearance on the L&WV was 21′5″ above the rail; the Erie insisted on a clearance of 27 feet. Murphy was willing to settle for a compromise figure of 23 feet. The Erie superintendent claimed this would present a danger to men working on the tops of cars. Murphy retorted that a man standing on a 15-foot boxcar would not hit the wire. He said the trolley wire carried only 600 volts and even a man swinging a lantern would receive only "a slight jar" if he hit the wire. Evidently Murphy was convincing, for after nine months of wrangling the Erie agreed to a height of 23 feet.

The line's few grade crossings presented another difficulty for the harried vice president of the L&WV. The growing number of automobiles had made the Public Service Commission increasingly conscious of grade crossing safety. In response to a survey, Murphy reported that the line had six crossings in Wilkes-Barre, two in Pittston, one in Avoca and four on the Dunmore branch. The crossings in Wilkes-Barre and Pittston were protected by watchmen, while the others ". . . were very infrequently used by pedestrians."

Like so many improvements made by the Laurel Line in the public interest, these illuminated crossing signals installed at South Street, South Avoca in August, 1929, were placed there not out of the railroad's generosity but because of outside pressure. There were many varieties of warning signals during that era before standardized configurations were mandated by government fiat.
Edward S. Miller

The following year the PSC demanded more information, particularly about the watchmen. Murphy wrote that they were on duty from 5:00 A.M. to 1:00 A.M. daily and that during the four hours in the early morning, the few cars running practically stopped at these intersections. The Avoca crossing had no watchman, but Murphy claimed since it was near the station where 90 percent of the cars stopped and had a good view, there was no need for one at this point.

Although the vice president did not tell the PSC, he was well aware that even where watchmen were on duty the intersections were still hazardous. He complained to the Wilkes-Barre chief of police that from April 19 to 30, a company watchman indicated 19 people ignored his signal to stop, including three trucks, despite warning blasts from the train whistle. One motorist ". . . very near drove over" the watchman. These hazards of the automobile age simply had to be tolerated by the L&WV. In 1929, under pressure from the PSC, the line did install an automatic warning signal at the Avoca grade crossing at a cost of $1,100.

In addition to careless motorists, Murphy had to deal with the results of carelessness by his maintenance workers. Individuals in Pittston complained of dirt washing down the fill at Railroad Street, which was damaging their property. No sooner was this corrected than some people in Moosic and the borough council complained the L&WV was dumping ashes and cinders into a small stream, blocking it and causing flooding on adjoining properties. This, too, had to be corrected.

Railroads in coal mining regions were perpetually waging war with mine subsidence and related problems. This view and that on the facing page, taken near Inkerman station in the spring of 1926, show the results of reckless strip mining too close to the right-of-way. A section crew is shoveling fill to restore the subgrade.
William F. Ellis Collection

The perennial problem of coal pillars under the right-of-way recurred in December 1923. The Hillside Coal and Iron Company presented a bill for $487 for coal under the line at North Pittston, and the Pennsylvania Coal Company claimed $1,211 for coal under the tracks between South Pittston and Hilldale. Murphy objected to the latter claim on the grounds that he had earlier requested the company to reserve the coal under the track near Inkerman for a distance of about 800 feet. The company agreed to do so and pillars were blocked out, but then a strike took place, disrupting mining operations.

When work resumed, the men apparently forgot their instructions and removed one of the pillars, thereby causing a cave-in, which derailed the 11:59 P.M. train from Wilkes-Barre on November 21. Six people were injured and the accident cost the line almost $3,000 with the possibility of more expenses due to a lawsuit from one of the passengers. Therefore, Murphy wanted an adjustment from the coal company on its bill to take into account the losses suffered by the Laurel Line as a result of the mining company's negligence.

Another accident had taken place in the tunnel shortly after 3:00 A.M. on January 3, 1920. A large piece of rock weighing about two tons fell from the roof directly onto the tracks at a point 390 feet from the south portal. Train number 1, northbound from Wilkes-Barre, struck the rock and knocked it out of the way. The car was derailed and the journal boxes and steps were damaged. Apparently, the rock had broken loose because of expansion and contraction from freezing water. No other sections of the tunnel were affected and normal operations resumed.

Disputes with the Scranton Electric Company. The Lackawanna and Wyoming Valley Power Company continued to produce electricity for its parent railroad and for sale to outside customers including the Scranton Electric Company. During the five-year period from 1919 through 1923, the power company earned a gross revenue of $1,282,642, an average of $256,528 per year. Of the total amount, $643,564 or 50 percent was received from the railroad company, $404,121 or 32 percent was earned from the Scranton Electric Company and $234,956 or 18 percent was paid by various local industries.

Because of this heavy usage, the railroad company felt the power plant was depreciating much more rapidly than expected. Originally, it was estimated the engines and turbines installed in 1903 and 1906 would have a life of 30 years, but it was now calculated that they would wear out in a considerably shorter period of time. Therefore, the railroad company wanted to arrange a supplementary contract whereby the power company would increase its rental payments to the extent that the railroad could fully depreciate its investment in the power facilities by 1930.

S.C. Stivers, secretary of the L&WV, anticipated by then a large power plant would be operating in Scranton from which the line could purchase all of its energy, thereby closing its own facility. These changes were internal, paper transactions which would benefit the railroad's financial situation. As it developed, the L&WV power company would have to struggle on until the end of electric operations in 1953.

Typical customers of the power company were the Spruks Brothers Lumber Company, McGregor and Peters Iron Works, the Glen Amusement Company, and various coal companies. These industries operated close to the right-of-way and could easily be served from the power company's transmission lines.

The L&WV continued to burn culm which, while cheap, sometimes caused pollution. The Scranton Dress Company complained to Murphy that dirt from the smokestack of the powerhouse was so bad they had to remove a cart load of it from their roof every six weeks. Murphy personally called at the dress company's offices to explain the problem.

As the supply of culm acquired in 1915 gradually diminished, the line began to look elsewhere for fuel. One possibility was to secure a lease of the Prospect culm bank of the Lehigh Valley Coal Company. As early as 1919, Ford and Murphy attempted to make this acquisition but internal problems within the corporate structure of the coal company prevented a formal agreement. Nonetheless, the L&WV was permitted to take the silt for its own use and had done so for a period of years. By 1927, the quality of the remaining silt was so poor it had to be mixed with coal to make it combustible. Murphy requested Sproul to persuade Lehigh Valley officials to sign the long-delayed lease so the L&WV could continue developing the site without danger of litigation or loss.

The contract with the Scranton Electric Company, signed in 1917, was a source of much trouble. On November 15, 1921, Murphy received a registered letter from the president of the utility, accusing the L&WV Power and Railroad Companies of violating the contract by failing to deliver electric energy as required for various periods over the past year. Unless these lapses were corrected and the price adjusted, he threatened to cancel the contract in 30 days.

A surprised Murphy attempted to explain the situation to Ford. He noted the L&WV power company had been forced to partially shut down one of its turbines for repairs in July. They had informed Scranton Electric of this and were prepared to supply 2,000 kw if the electric company required it. The utility did not, in fact, request any power, but it deducted from the L&WV bills a demand charge for the days when the power plant was below capacity. Murphy argued that since the L&WV was prepared to furnish some power, if necessary, the demand clause of the contract was met and the L&WV should be paid in full.

Ford pointed out the electric company was installing a new 25,000-kw unit, which would lower its cost significantly and lessen the need for power from the L&WV. He asked the vice president to determine whether the Laurel Line could sell its power to new customers such as coal companies in the event the contract were canceled. Murphy replied that if the contract were abrogated the L&WV would lose about $20,000. This might be offset, however, by selling surplus power to potential new customers in the vicinity of Scranton. He charged the Scranton Electric Company with attempting ". . . to do us as much harm as they possibly can."

Murphy feared when the new generator, which Scranton Electric was installing, became operational, it would cease to purchase energy from the L&WV during the daytime but would still insist on selling electricity to the railway during the nighttime hours when the Laurel Line's plant was closed.

Ford wanted to remain on good terms with the utility and he arranged a series of conferences in December between officials of the Laurel Line and Scranton Electric. Several suggestions came from these meetings. It was proposed the contract be retained and the electric company should pay the amounts it had previously deducted. The L&WV would reduce its capacity reserved for the electric company from 3,000 to 2,000 kw with a corresponding reduction in the fixed annual payment to it from $30,000 to $20,000. Effective January 1, the L&WV would generate its own power from 11:00 P.M. to 7:00 A.M. except during emergencies. An agreement was drawn up along these lines, despite the objections of Murphy, who felt the electric company had been taking advantage of the Laurel Line and would continue to do so. Ford was anxious for a settlement and overruled the vice president and the supplementary agreement to the contract went into effect in February 1922.

The Laurel Line Embraces the Bus. The development of paved highways provided both a threat and an opportunity for the Laurel Line. Competition could be expected in the form of buses and trucks operated by other individuals or carriers. To an extent, the L&WV might counteract this danger by establishing its own lines on the public highways. The first serious consideration of a bus extension occurred in 1921 when Ford asked Murphy if the company could increase its business by establishing motor bus extensions to such towns as Carbondale, Kingston and Nanticoke. The vice president was not enthusiastic about this proposal. He felt the Public Service Commission would deny the L&WV permission to establish bus lines because these towns were already served by the Wilkes-Barre Railway, the Scranton Railway Company and some of the steam railroads.

Although extending its operations to more distant off-line points was not feasible, the company decided to begin a bus service between Moosic and Pittston. The Scranton Railway Company had applied to the PSC in 1923 for permission to abandon its streetcar line from Scranton to Pittston. The approval of this application gave the L&WV an opportunity to establish its own route in this area.

Initially, an organization known as the Laurel Line Bus Company was incorporated in 1924 by Murphy and several other individuals. It was authorized to operate a bus route in the territory west of the Laurel Line tracks. Beginning at the Moosic station of the L&WV, the line ran along Springbrook Avenue to state highway Route 5, and along that road to Main Street, which it followed through the boroughs of Avoca and Dupont to Pittston, where it terminated at the corner of Main and Broad Streets, a distance of about five and one-half miles. Service began at the end of June 1924. Four 25-passenger buses were purchased and fares ranged from eight to 16 cents.

The operation was considered sufficiently successful so that in 1925 Murphy formally sold his bus company shares to the L&WV Railroad for $20,000. Fourteen thousand dollars of this he received in the form of 560 shares of L&WV stock, valued at $25 a share. The company amended its charter in January 1928 and received the power to operate anywhere in Lackawanna and Luzerne Counties and to offer charter service through the state.

The bus fleet was slowly expanded. As of 1928, there were five vehicles numbered 15 through 19, mostly built by Mack, at an average cost of $6,200. One bus was equipped with a snowplow with an eight-foot blade. Although the railroad ran its entire life without the benefit of a snowplow, evidently the buses were unable to make it through the first winter without one. Eight drivers were employed on the route and a garage was acquired in Avoca.

Financially, the bus operation reported slight profits. For example, in 1927 the line grossed $56,000 and had operating expenses of $50,000. When taxes and other deductions were made, the company reported a net income of only $2,400. By 1929, revenue had fallen to $51,000, but expenses remained essentially the same and a base profit of $369 was realized. An average of about 580,000 passengers were carried per year, but the low fares kept revenues down.

This situation prompted Murphy to complain to the city of Pittston that its tax levy of $150 per year was excessive. He told the city officials that only two buses ran in the city at any one time and added, "We are losing money operating three buses and we are trying to get . . . the bus company to pay its own way." The next decade would see an expansion of highway operations but profits would still be minimal.

Two separate local trolley routes once connected Scranton with Wilkes-Barre but they hardly provided competition to the swift cars of the Laurel Line. Wilkes-Barre Railway Corp. 350 (ABOVE), built in 1911 but very similar to cars erected in 1902, awaits departure time back to the county seat from Duryea on September 7, 1932. That same morning, Scranton Railway Co. 355 (BELOW) had just arrived at the end of the track in Old Forge. The Scranton Railway tracks extending from Old Forge to the Wilkes-Barre Railway connection in Duryea were abandoned in 1926. Two years previously, the abandonment of the Scranton Railway line to Pittston provided the Laurel Line with an opportunity to establish a bus subsidiary. *BOTH: William H. Watts, negatives in Francis J. Goldsmith, Jr. Collection*

New Steel Cars Arrive. The decade of the 1920s would see significant changes in the equipment operated by the Laurel Line. Locomotive 402, which had been damaged in several earlier accidents, was rebuilt, except for its trucks. It received a new frame, body repairs, and new wiring and it was returned to service in November 1921. Locomotive 403 was equipped with new Westinghouse 577 motors in January 1922. Murphy had considered buying General Electric 207D 165-hp motors for this engine, but the price of $2,275 per motor he considered as "exorbitantly high."

The passenger cars were modified by the application of solid steel rolled wheels instead of the steel tire wheels used previously. A set of wheels purchased from the Carnegie Steel Company was tested for a year and judged to be superior to the earlier model. Additional sets were purchased and gradually placed on all the cars.

The original type L cars were deteriorating and replacements were in order. In typical fashion, the Laurel Line management sought an economical method of securing new equipment. Beginning in 1924, the line purchased steel car bodies from the Osgood Bradley plant in Worcester, Massachusetts. Trucks, motors and controls were taken from older cars about to be scrapped and placed under the new bodies. In this way, the line was able to modernize its fleet at a minimal cost.

Coaches 31–34 were acquired in 1924 at a cost of $27,000 each. In the following year, 35 and 36 were received for the same price. In 1928 and 1929, the remaining coaches, numbers 37, 38 and 39, arrived from the builder at a price of $21,000 each. These cars, and combination car 116, were equipped with Hyatt roller bearings. They were also 2′8″ longer and slightly heavier, caused by the addition of an extra window panel. Combines 114 and 115 were purchased in 1925 and 116 entered service in 1929.

The new equipment, like the old, had two Westinghouse 140-hp motors, with Baldwin trucks. Motor truck wheels were 36″ in diameter and trailer truck wheels were 33″. The cars were 60 feet long, nine feet wide, had a truck wheelbase of seven feet and weighed an impressive 80,000 pounds. Control

The Laurel Line purchased only the steel car bodies, including interior appointments, from Osgood Bradley. The completed bodies of 33, built in 1924, and 35, in 1925, were resting temporarily on shop trucks to allow movement around the factory complex in Worcester, Mass. The collapsible seats can be seen in the pristine view of the baggage compartment of one of the combines. Control apparatus, motors, gears, trucks, air-brake equipment, and even trolley poles were salvaged from obsolete type M cars and installed on the new cars in Scranton. The first order came without couplers — the Laurel Line had some in stock. The second order had couplers fitted at the factory. Note the extra wide window post marking the division of the smoking and non-smoking sections. The 1928 steel cars had one extra window. *George Krambles Collection*

Both dimensionally and stylistically, the Osgood Bradley-built steel cars helped to reinforce the carrier's claim that it was a railroad, not an interurban. The reversible low-back seats, luggage racks, window shades, and lamp fixtures were typical of short-distance railroad day coaches of that period. One-third of the car was designated as a smoking compartment and was fitted with leather seats. The non-smoking section had green plush upholstery. Mahogany was used for the partition between the two sections and for trim. One railroad car feature was missing: the Laurel Line was short enough that toilets were an unneeded luxury.
George Krambles Collection

was of the Westinghouse double-end AB, multiple-unit type with circuits carried through Tomlinson couplers. Balancing speed was about 50 mph. The interior was divided into passenger and smoker sections and was finished in varnished wood. The passenger section had reversible green plush seats and the smoker was equipped with imitation leather seats. All were fitted with baggage racks and sliding doors in the bulkheads.

As the new cars entered service, most of the old wooden-type L coaches were scrapped. By April 1925, only numbers 8 and 10 from this series were still on the roster and, of the combines, only 104 remained active. Number 101 was converted to a baggage trailer by 1927. Of the four express cars, 201 and 203 were gone from the property by 1926.

About this time, a new third rail sleet scraper was developed in the company's shop. The device was air operated and attached to the shoe beam on the trucks. It consisted of steel scraper blades, which could be lowered by the motorman to remove ice and snow from the third rail. The blades gave about 200 miles of service. This improvement made it unnecessary to send out specially equipped cars when ice and snowstorms hit the valley.

There were minimal changes in the freight car fleet. As of 1929, the company still had nine boxcars in the 500 series; nine flatcars, numbered in the 600 and 1000 series, one of which, 1005, was an idler for the derrick; 15 steel hoppers in the 751 series and eight ash cars in the 711 series. Wear and tear would gradually reduce the number of freight cars.

During the late 1920s and 1930s, the Laurel Line mixed wood cars into trains with its new steel car fleet when traffic demanded. The older equipment could operate in multiple with the newer equipment because they had identical electrical and air brake apparatus. This view shows a type M car behind two steel cars on the Avoca Viaduct in 1939, near the end of such service. The automobiles outside the Heidelberg colliery no doubt belonged to the mine boss and his supervisory staff.
Michael J. Lavelle, Edward S. Miller Collection

Financial Status. The decade of the 1920s saw the Laurel Line reach its peak in terms of earnings and profits. It also witnessed the beginning of the carrier's plunge into red ink, which would characterize the line in the Depression period of the 1930s.

The same management team remained in control. In 1921, officers included President Sproul of Chester; First Vice President and General Manager Murphy of Scranton; Secretary S.C. Stivers of New York; Treasurer F.W. Potts of New York; and Superintendent J.H. Kilker of Scranton. The Executive Committee included Ford, Sproul and Walter Lippincott.

Also continuing from previous years was the line's tendency to delay paying its bills. Auditor W.E. Pearson received a pleading letter from the Philadelphia office of the Westinghouse Electric and Manufacturing Company in May 1921, asking for the prompt remission of $2,270 for invoices which went back to May 1918. A Westinghouse official, F.P. Bell, complained he had sent numerous statements asking for settlement and confessed he was ". . . very much discouraged at the necessity of continued requests on our part for settlement of the unpaid charges." He added that if the line paid he could dispose of a file of correspondence which had accumulated over the past two years. Much to his relief, the L&WV finally met its overdue obligations.

Ford, Bacon and Davis continued to serve as management consultants and effectively had the final word on corporate decisions. In general, Ford still exuded optimism about the carrier's prospects. Indeed, he even considered acquiring additional interests. In September 1921, he met with S.Z. Mitchell of the United Gas and Electric Corporation and discussed the possibility of acquiring the Lehigh Valley Transit Company, which operated the electric railway system in the Allentown–Bethlehem–Easton area. In common with many utilities of the day, UG&E was busy merging its various operations into holding companies and was anxious to dispose of its only railway operation, the LVT. No agreement was reached, since Ford was willing to pay only 50 percent of the physical value of the traction company. As it developed, the Lehigh Valley Transit Company came under the control of the Electric Bond and Share Company.

Had Frank Ford been successful in his overture to acquire the Lehigh Valley Transit Co., the interurban line in the Wyoming Valley would have been dwarfed by a subsidiary almost four times its size. The LVT operated almost 250 cars over 164 miles of track connecting suburban Philadelphia with Allentown, Bethlehem, Easton, Nazareth, Slatington, Slatedale, Macungie, Hellertown, and Phillipsburg and Alpha, New Jersey. During the Allentown fair in September, 1920, LVT ran everything capable of turning a wheel. Open cars built in the 19th century were sharing the streets with center-entrance steel cars only one year old. The photograph was exposed at 6th and Hamilton Streets in Allentown by Frederick E. Barber, who began aiming a box camera at trolleys as early as 1912.

LVT's limited services to Philadelphia were provided by heavy-weight interurban cars built to Middle-western standards, such as this train of 1912 Jewett-built cars (ABOVE) on Allentown's South 8th Street viaduct. *George Krambles Collection* **The Easton Division lines focused on attractive Center Square (RIGHT), at the intersection of 3rd and Northampton Streets. Single truck cars were the mainstay of Easton services until 1939.** *Alan Martlew Collection*

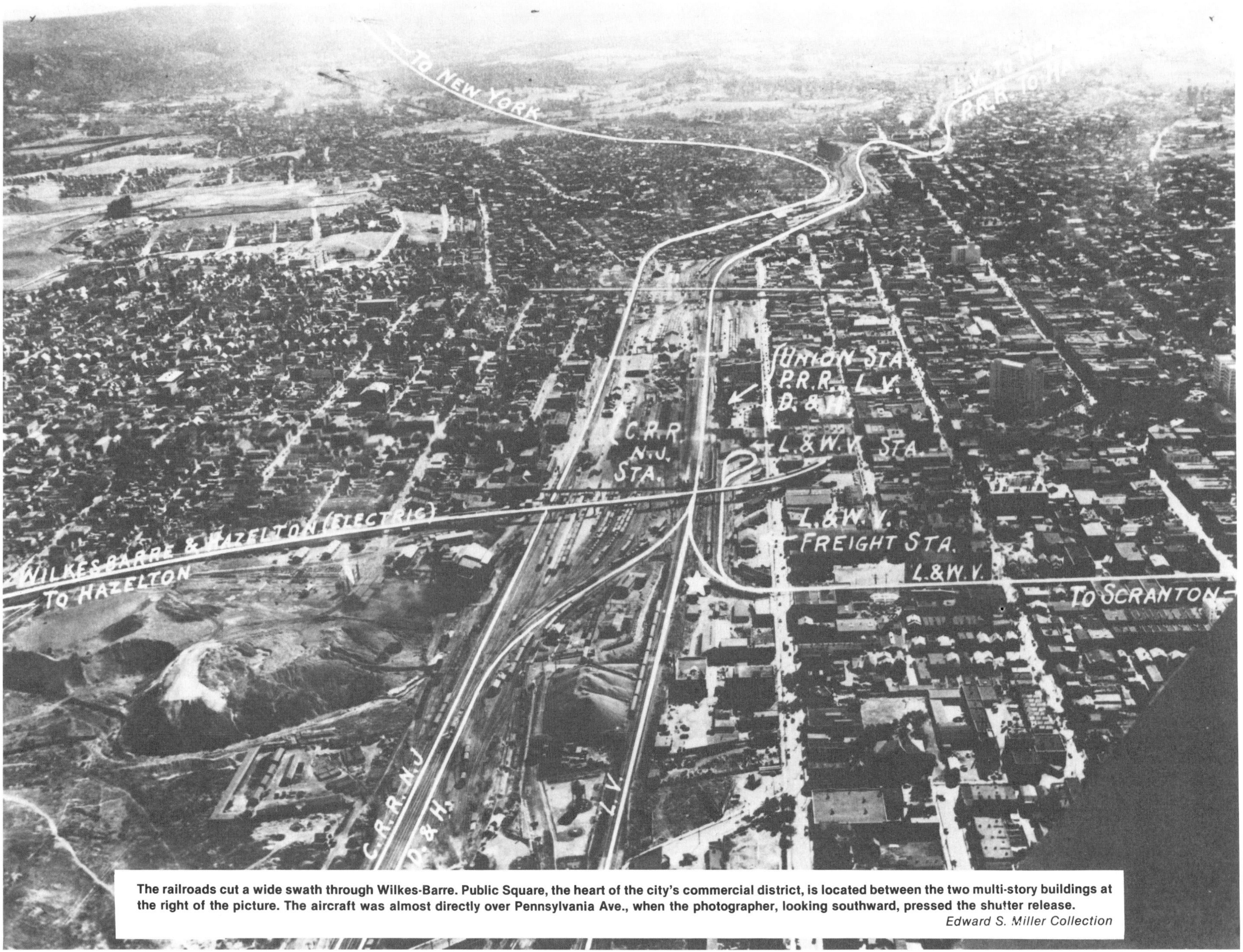

The railroads cut a wide swath through Wilkes-Barre. Public Square, the heart of the city's commercial district, is located between the two multi-story buildings at the right of the picture. The aircraft was almost directly over Pennsylvania Ave., when the photographer, looking southward, pressed the shutter release.

Edward S. Miller Collection

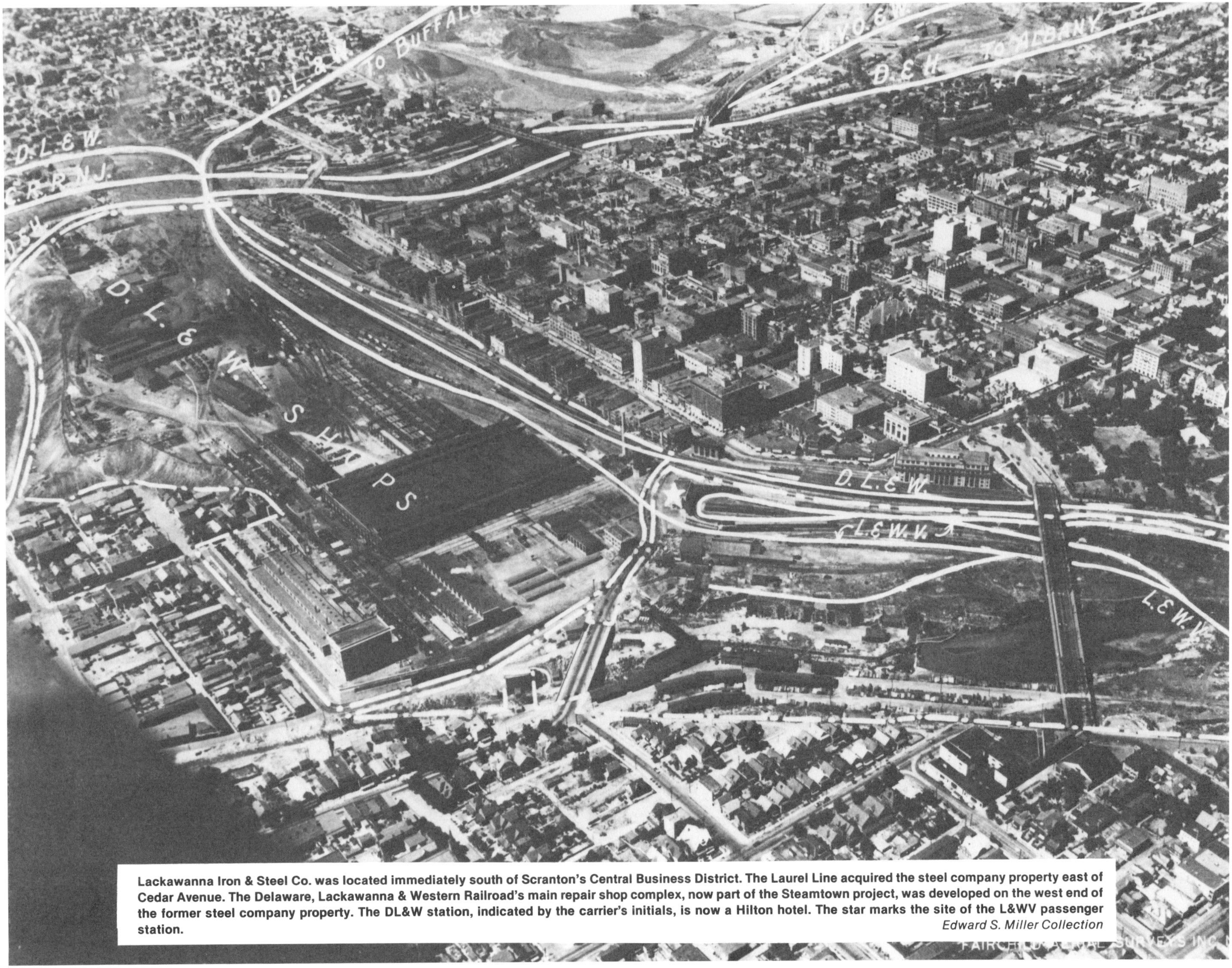

Lackawanna Iron & Steel Co. was located immediately south of Scranton's Central Business District. The Laurel Line acquired the steel company property east of Cedar Avenue. The Delaware, Lackawanna & Western Railroad's main repair shop complex, now part of the Steamtown project, was developed on the west end of the former steel company property. The DL&W station, indicated by the carrier's initials, is now a Hilton hotel. The star marks the site of the L&WV passenger station.
Edward S. Miller Collection

Two of the original wooden cars built for the Philadelphia & Western Railway rot on the rip track behind the 72nd Street Shops in Upper Darby, Pa. in 1934. These cars were in active service in 1921 when Frank Ford ordered a cost-of-operation study comparing the L&WV and the P&W. *Francis J. Goldsmith, Jr.*

Ford was also interested in the Wilkes-Barre Railway Company, which he felt might be available after the sudden death of T.A. Wright, vice president and general manager, in an automobile accident. He wrote to F.M. Kirby, president of the Wilkes-Barre Railway, suggesting he hire Ford, Bacon and Davis to manage the property. Although Ford argued that his firm's experience with the Laurel Line made him aware of the area's potential and dangled the possibility that he might buy into the company, no agreement was reached.

Ford made a study in 1921 comparing the expenses of the Philadelphia and Western Railway and the Laurel Line. The P&W, as it was known, also was a third rail line providing service between Upper Darby and Norristown, a distance of 13 miles, with a branch to Strafford. The study showed the Laurel Line expenditures for maintenance, conducting transportation and general costs were considerably higher than its neighbor to the south. Total operating expenses for the L&WV were $867,000 and for the P&W only $487,000. Net income was $378,000 and $145,000 respectively. It would seem that overhead and payroll expenditures were considerably higher for the Laurel Line. No action was taken as a result of the study. Aside from their method of propulsion and private right-of-way, the two properties were quite different in such respects as freight, crew size, and type of service offered. Not surprisingly, no conclusions were drawn as a result of the study.

As indicated in the chart following, 1925 saw the line hit its peak earnings. A decline began in the following year, although net income was greater than previously. The real slide in revenue passengers and total earnings began in 1927 as concrete highways further penetrated the area. The last sections of a through highway were constructed in Moosic, Avoca, and Dupont in 1928 and immediately automobile and bus competition made itself felt. Some 50,000 fewer passengers were carried in that year than in 1927. By 1930, a further decrease of 700,000 riders was recorded.

Altogether, in the decade of the 1920s, revenue passengers declined 48 percent, a tremendous loss for a small carrier as heavily oriented as the Laurel Line was toward the passenger trade. In similar fashion, ticket sales to Rocky Glen Park fell 65 percent during the same period. Freight revenues also declined but at the less dramatic rate of 33 percent. The operating ratio, while still favorable, rose from 64 percent to 75 percent between 1925 and 1930. In the latter year, the company recorded the first of what was to be an almost endless series of deficits.

Financial Reorganization—1928. Declining revenues and falling traffic indicated that difficult days were ahead for the Laurel Line. Ford, Bacon and Davis concluded that the situation would be improved by a financial reorganization, which would eliminate the holding company, the Scranton and Wilkes-Barre Traction Corporation, and leave the Lackawanna and Wyoming Valley Railroad Company as the surviving firm. This would necessitate extensive legal steps and stockholder approval, as well as the issuing of new securities by the L&WV to acquire and retire the stocks and bonds of the S&WB Traction Corporation. In a letter to the stockholders in January 1928, Ford, Bacon and Davis noted the S&WB Traction Corporation had issued $4 million in five percent and six percent bonds, $1 million of seven percent preferred stock and $2,400,000 of common stock.

The management firm argued that traction securities in general had become unpopular, with a very small market for trading. Ford, Bacon and Davis also pointed out that a new federal income tax law would increase expenses for the separate companies. They therefore proposed the traction corporation should redeem its bonds and exchange stock with the L&WV Railroad. As the traction stock was listed at $100 par value and the railroad stock at $50 par value, each holder would receive two shares of the latter for one of the former.

To accomplish this, the railroad company issued $2,900,000 first mortgage five percent bonds, which were discounted to 89 percent each and $1,110,000 of 20-year six percent debentures, discounted to 90 percent each. From this sale the company realized $3,571,000. This sum, plus 20,000 shares of preferred stock valued at $1 million and 5,000 shares of common stock valued at $250,000, were turned over to the traction corporation in exchange for the $5 million five percent first mortgage gold bonds held by the company. In effect, the L&WV redeemed its 1913 five percent gold bonds, which were worth $5 million, for a payment of $4,821,000.

As part of this financial maneuvering, the traction corporation was able to retire its obligations and was formally dissolved by the Commonwealth of Virginia on December 5, 1928. Most of the stockholders were content to receive the new L&WV shares, a move which they soon regretted, but a few shrewd individuals insisted on being paid off in cash and thus escaped the oncoming financial collapse of the Laurel Line. The L&WV also received the entire outstanding capital stock of 1,200 shares of the Laurel Line Bus Company, which had been owned by the traction corporation.

As part of the general belt tightening, the contract with Ford, Bacon and Davis was not renewed after 1930. For better or worse, the future of the L&WV rested in the hands of Patrick J. Murphy, who became president of the corporation. This was a logical step in that the engineering firm was no longer interested in traction companies and with its greatly reduced prospects the Laurel Line no longer had need for outside consultants.

CONSOLIDATED INCOME STATEMENT

Lackawanna and Wyoming Valley Railroad Company

	1921	1922	1923	1924	1925	1926	1927	1928	1929	1930
REVENUE										
Passenger	1,067,950	940,544	1,019,406	1,030,188	974,839	865,024	849,558	738,869	631,336	532,619
Freight	186,240	196,224	257,901	218,002	164,638	173,827	148,925	138,510	121,970	111,185
Other	209,101	163,634	171,430	162,285	45,746	40,818	41,839	41,622	39,980	38,975
Total	1,463,296	1,299,802	1,448,738	1,410,477	1,185,255	1,079,701	1,045,856	924,869	793,286	688,512
EXPENSES										
Maintenance of Way & Structure	187,803	141,437	194,084	200,977	141,753	113,467	92,278	96,371	88,693	65,620
Maintenance of Equipment	126,374	101,671	132,243	127,813	112,356	96,004	75,849	67,926	62,531	58,426
Power	249,755	194,245	184,068	186,665	165,872	155,977	141,560	134,392	118,923	124,585
Transportation	243,221	234,189	249,966	245,098	236,094	231,216	222,756	215,560	201,893	189,503
Other	167,275	138,749	133,031	129,196	105,956	109,850	96,097	106,169	105,610	80,647
Total	974,430	810,182	893,393	889,750	762,031	706,515	628,541	620,419	577,650	518,784
NET REVENUE	488,866	489,620	555,345	520,727	423,194	373,186	417,314	304,450	221,630	169,729
Taxes	39,520	51,539	56,800	61,279	51,514	36,499	37,874	40,727	20,794	13,945
GROSS INCOME	403,063	390,251	451,071	409,852	397,610	455,182	420,039	301,365	273,059	207,326
Interest	205,842	204,017	202,169	201,063	250,475	250,000	250,000	197,654	222,270	222,144
NET INCOME	197,221	186,235	248,902	208,788	147,135	205,182	170,039	102,956	48,772	(−16,994)
DIVIDENDS PAID	208,391	154,719	181,946	181,931	174,150	166,625	166,625	49,592	65,486	16,356
SURPLUS	(−11,170)	31,516	66,956	26,857	(−27,015)	38,557	3,414	53,364	(−16,714)	(−33,350)

Source: Financial Report, Before the National Railway Labor Panel Emergency Board, 1925-1942, Miller Collection.

Miscellaneous Statistics

	1925	1926	1927	1928	1929	1930
Passenger Car Mileage	1,292,873	1,292,942	1,301,561	1,197,028	986,200	925,283
Freight Car Mileage	233,693	244,742	239,876	244,447	228,885	230,314
Total Car Mileage	1,526,566	1,517,684	1,541,437	1,441,475	1,215,085	1,155,597
Revenue Passengers Carried	3,932,969	3,665,960	3,389,779	2,849,242	2,424,625	2,108,794
Passenger Revenue	$974,839	$865,024	$849,558	$738,870	$631,336	$532,619
Average Fare	.248	.236	.251	.26	.26	.253
Total Operating Revenues	$1,185,225	$1,079,701	$1,045,856	$924,869	$799,280	$688,512
Operating Revenue per Car Mile	.776	.717	.678	.641	.658	.596
Total Operating Expenses	$762,031	$706,515	$628,541	$620,419	$577,650	$518,784
Operating Expense per Car Mile	.50	.466	.408	.43	.475	.449

Source: Report before the National Railway Labor Panel Emergency Board, 1925-1942, Miller Collection.

Squealing around the tight curve near Laurel Junction, one-man wooden car 19 reflects depression era economies.
Edward S. Miller

5

Through the Depression and War with the Laurel Line

EVEN BEFORE MURPHY assumed control of the L&WV, the superficial prosperity of the twenties came to an end. The signal which touched off the Great Depression of the 1930s was the collapse of the stock market on October 29, 1929, when more than 16,400,000 shares were sold for whatever price they would bring. There were many complex factors that, combined, resulted in the economic disaster of the 1930s, but for the Laurel Line the main problems were drastic declines in passenger traffic and revenue, with even larger deficits as a result.

Republican President Herbert Hoover, who was inaugurated on March 4, 1929, had to bear the brunt of the blame for the staggering economy. Hoover was a humane man, but, as a classic "rugged individualist," he felt the state and local governments, businesses and private charities had to carry the major burden of relief and revitalization. Although he took several daring steps, such as supporting agricultural reform and the creation of the Reconstruction Finance Corporation, basically, the Hoover administration was overwhelmed by the magnitude of the Depression. In the 1932 elections, the Democrats, led by Franklin Delano Roosevelt, scored a decisive victory and returned once more to power.

Roosevelt's policies, referred to as the New Deal, were essentially pragmatic, but they wrought major social, political and economic changes in American life. The federal government assumed tremendous powers and created dozens of new agencies to combat the widespread suffering and misery. Despite these heroic measures, the Depression persisted, and not until the start of World War II in Europe in 1939 did the American economy significantly recover. For the Laurel Line, the improved economy came just in time and helped to prolong the life of the carrier for another two decades.

The L&WV Soldiers On. The most striking and alarming feature of the Depression on the Laurel Line was the drastic and seemingly endless fall in passenger traffic. At the start of the decade in 1930, 2,108,794 fares were collected for a total revenue of $532,619. By 1932, generally considered the worst year of the Depression, the passenger count declined to 1,286,793, producing $296,243. Despite New Deal relief and recovery efforts, 1937 saw the line carry 1,187,384 people for a revenue of $230,834. The worst year for passenger traffic was 1940, after the economy had begun to recover. Despite this, the line hauled only 1,036,010 passengers, who paid $190,252 in fares. Over the 10-year period, therefore, passengers carried decreased by 51 percent and revenue fell 64 percent.

The management struggled valiantly but ineffectively to deal with this situation. Fares were reduced in an effort to attract more riders. On May 1, 1932, round-trip fares were cut 15 percent. One-way fares were dropped about 12 percent on June 1, 1936. At the same time, the company finally took official cognizance of the fact that all passenger trains ran through the Scranton tunnel and slightly reduced 10-trip fares because of the shortened mileage. Early in 1940, the 10-trip ticket was further slashed by 20 percent.

Surprisingly, schedules were not drastically changed. In May 1931, eight weekday limited trains were removed from the time card, ending 20-minute headways, except during rush hours. A similar number of runs were eliminated from the Sunday schedule. No further changes were effected until December 1939, when half-hourly service was established from 6:00 P.M. to midnight. Most trains required only single cars because of the light loads.

Efforts were made to encourage business by running excursions, but these, too, showed the same downward trend. For example, in 1930, 146 excursions were operated, carrying 5,745 passengers and earning $4,126. By 1932, 138 trips were run carrying 6,077 people who paid $3,145. Personally conducted tours of Washington, D.C., were offered in conjunction with the Pennsylvania Railroad. In 1930 and in 1932, four such tours were arranged, but passengers carried fell from 397 to 156 and revenue declined from $478 to $156. The company attributed this decrease to competition by the Martz Coach and Eastern Greyhound Lines, both of which were actively soliciting customers and offering more flexible schedules as well as a one-seat ride.

Another attempt to stimulate traffic was the running of New York City excursions in connection with the Lehigh Valley Railroad. The fare was an amazingly low $2.50 round

Watching Trains in Wilkes-Barre

Wilkes-Barre Union Station (LEFT), located on the south side of E. Market Street across from the L&WV station, was home to three of the five railroads serving that city and nearby Kingston. Some 42 passenger trains stopped here each day when the area's population peaked in 1930. Two of the railroads advertised connecting trains of the L&WV in their timetables. The one that did not was the Delaware & Hudson, which scheduled 11 daily trains in competition to the L&WV, two of which offered connections to and from Albany. One of those Wilkes-Barre to Carbondale locals (BELOW) awaits departure time in 1938, three years before the service expired.

LEFT: Edward S. Miller Collection; BELOW: Michael J. Lavelle

When the stock market crashed, the Pennsylvania Railroad carded two daily round-trips from Wilkes-Barre to Philadelphia via Hazleton, Pottsville and Reading. An additional four round-trips to Sunbury connected there for Baltimore, Washington, Harrisburg, Buffalo, Elmira, Erie and Pittsburgh. A spartan train for Sunbury (ABOVE) will leave behind a hand-fired freight engine in the late 1930's. The Lehigh Valley Railroad, owner of the station, offered the most extensive service because Wilkes-Barre was astride its main line from Buffalo to New York; its trains carried through cars for Chicago and Toronto. The Valley's "John Wilkes," streamlined in 1939 for the daylight trade to Gotham, pauses at Pittston at the start of its journey. Its next stop would be Wilkes-Barre. The last passenger train to call at Wilkes-Barre was the Lehigh Valley's "Maple Leaf" for Buffalo and Toronto on February 4, 1961. *BOTH: Michael J. Lavelle, Edward S. Miller Collection*

The usual group of idlers are congregating on the sidewalk in front of the rather austere Wilkes-Barre terminal around 1930. Inside the terminal, workers were busy removing years of accumulated grime. By the end of World War II, the stations were again shabby and run down, prompting E. McLain Watters to order another cleanup.
BOTH: Edward S. Miller Collection

trip. Several options were offered, but only a small number of passengers availed themselves of the L&WV connecting trains.

The Depression also affected adversely ticket sales and revenue from service to Rocky Glen. Whereas in the early and mid-1920s the line sold an average of about 190,000 fares annually to the park, earning an average of approximately $60,000 in revenue per year, by 1933 ticket sales fell to less than 35,000 and only $9,600 was realized. Various promotions were undertaken by the park management and the Laurel Line shared, to a certain extent, from the crowds which were attracted to them. In August 1933, for example, Father Charles Coughlin, the famous (and soon to be controversial) radio priest from Royal Oak, Michigan, spoke at an Irish Day program at the park. The L&WV ran special trains for this event.

Certain days each season were proclaimed "Nickel Days," when most of the attractions were available for only five cents. Service clubs and other organizations were contacted and encouraged to use the park facilities for picnics and clambakes. In one week in July 1933 more than 700 circulars advertising the park were distributed to stores and outlets in Scranton and Wilkes-Barre.

The traditional summer holidays, Memorial Day, July 4th and Labor Day, always attracted additional riders. Despite these strenuous efforts, the trend was downward on the Laurel Line. Not only were hard times limiting the number of pleasure seekers, but many who did go to Rocky Glen preferred to use their own automobiles. The last half of the decade saw a slight upturn with annual revenues of about $10,500.

It seemed that the L&WV could not gain additional riders despite schedule and service changes made by its competitors. In the spring of 1933, the DL&W removed most of its passenger trains from the Bloomsburg division. Only two trains in each direction remained on the time card. Murphy estimated the L&WV would gain 50 to 60 passengers daily as a result. Included, however, were 35 or 40 employees of the DL&W, who could ride on the Laurel Line using half-fare commutation tickets. However, the next year the Scranton Bus Company began a new line to the southside section of the city. A four-day survey revealed the loss of 50 to 60 daily passengers for the L&WV with an even greater loss anticipated in the future.

To partly offset this decline in traffic, the L&WV gradually trimmed the ranks of its employees. In 1930, the carrier listed four general officers, eight clerks, 18 maintenance workers, 90 operating employees and 40 shop and other workers, for a total of 160. Ten years later, there were three general officers, six clerks, 16 maintenance workers, 76 operating employees and 27 others, totaling 128, a 20 percent reduction. Only 25 years earlier, the company employed more than 360 workers. Some savings were effected by closing facilities. For instance, the Pittston freight station was shut on May 16, 1931.

Wage rates were cut also in the continuing battle to keep the line alive. In 1930, passenger conductors and motormen earned 72¢ an hour, freight conductors and motormen 76¢ an hour and brakemen 65½¢, with the average weekly wage amounting to $40.32. Murphy, in contrast, received an annual salary of $11,880. On the roster were 37 regular trainmen, seven freight and extra men, and 10 brakemen; 50 of these 54 workers were union members. Regular workers were guaranteed eight hours' pay, with overtime beyond that point, and extra men received a minimum of four hours' pay. Trainmen spent about 10 percent of their time on layovers, and they were allowed 10 minutes before and after duty for reporting and making out trip sheets. The company advanced $10 in change to each conductor.

Motorman Matthias Loch and conductor Charles Dorsey pose in timeless fashion beside their car at the Wilkes-Barre loop in 1940. Their seniority dating to 1906 and 1907, respectively, helped them to weather the lean years. *Francis P. Kutta*

With red ink appearing in ever-growing volume on the ledger, the company tried to economize in 1932. Reduced schedules put into effect the previous year cut the number of regular workers slightly and the average weekly wage fell to $36. Around the same time, a 10 percent wage cut was implemented for both clerical and union help. In addition, one-man service was initiated on the Dunmore line, using cars 19 and 23. The cars were not actually converted until January 1938 when steps were removed from one side and controls shifted from the left to the right side of the cab.

The union protested these changes and invoked the arbitration provisions of the contract. W.L. Reed, vice president of the Brotherhood of Railroad Trainmen, came from Cleveland to testify. He argued at the hearing that the L&WV had earned "enormous profits" after 1913 but made no effort to share them with the men. He held it was inconsistent for the company now to expect the workers to shoulder some of the mounting losses and protested the use of one-man cars on the Dunmore branch.

Murphy responded by pointing out the extent of bus and truck competition and warned the company would abandon the Dunmore branch if it were forced to operate these cars with a two-man crew. He said the non-union employees had absorbed a 10 percent cut and it was only fair for the operating employees to do so as well.

Some 33,000 miners lost their jobs in Lackawanna and Luzerne counties between 1930 and 1935. Those who were still employed at mid-decade could count on only 180 to 200 days of work that year. After removing the last limited trains in 1931, L&WV management persisted in running local trains every 20 minutes in the peak hours and every 30 minutes otherwise. Considering the severity of the Depression in the area and the intensity of service still provided, two and three-car trains had become rare. The single combine, illustrated west of the shops in Scranton in 1935, was adequate. By the end of the Depression, schedules were finally cut in the rush hours to every 30 minutes. *Francis J. Goldsmith, Jr.*

The arbitration board ruled trainmen must accept the wage reduction and one-man service would continue on the Dunmore line. This decision remained in effect throughout the Depression, and the men counted themselves lucky to have any job, considering the state of the economy and the financial condition of the road. When the financial situation improved slightly in 1940, due to increased freight traffic, the union pressed for a restoration of the wage cut and an additional 5½ percent increase. Once more, the question was referred to an arbitration board, which ruled against the union on the grounds that the company's earnings did not warrant a raise and that the pay scales compared favorably with those of other interurban roads.

Accidents Continue on a Reduced Scale. The electric line continued to be a hazardous site for careless or thoughtless individuals. An average of three people per year were killed and 35 injured. Many of these accidents were caused by people stumbling on or ignoring the dangerous third rail. For instance, two young men left Sussex, New Jersey, in August 1935, seeking employment in Buffalo, New York. They hopped an Erie Railroad freight, which took them to Dunmore. Here, they walked south along the Laurel Line tracks, hoping to catch a Lackawanna freight to take them to the shore of Lake Erie. Evidently not realizing the third rail was charged with more than 600 volts, one of the men sat down on the track to remove his shoes. In a short time, he was dead. Even employees who were familiar with the danger sometimes were killed. In 1933, a shop worker who was repairing the brake rigging on a locomotive failed to realize a short circuit had charged a brake shoe. He was electrocuted when he came in contact with it.

Another tragedy occurred in August 1933. A former employee and union activist who was then living in Chicago had returned to Scranton to visit his old friends. While conversing with crew members on a train heading for Rocky Glen, he lost his balance, fell off the platform and was killed.

The automobile was not only siphoning off the line's passengers, but it was proving to be a hazard at the few grade crossings as well. An average of three to five collisions a year occurred, mostly in Wilkes-Barre, when hasty drivers ignored the crossing guard's warnings and drove directly in front of oncoming trains. Injuries and sometimes fatalities were the sad result. Even people planning suicides sometimes chose the Laurel Line as the means of ending their earthly existence. One unfortunate individual in 1938 brought needless notoriety to the L&WV by first slashing his throat with a bread knife and then, soaked with blood, falling across the Laurel Line tracks in Plains where, a few minutes later, a speeding train mangled his body.

There were the usual minor derailments of the type to be expected on any railroad. One occurred in August 1939 as locomotive 403, heading north with a train of six cars, rounded a curve near Pittston. As it did so, the journal bearings on a refrigerator car became dislodged, causing one truck to derail. Fortunately, the train was moving slowly, and only the northbound track was affected until the errant car was returned to the rails.

Right-of-Way and Equipment Changes. As was to be expected from a line as hard-pressed financially as the L&WV had become, no major improvements were undertaken during the Depression years. Minor changes that were made were done with an eye to reducing labor and operating costs. Several tons of new rail were installed in 1930. The Rocky Glen pavilion was rebuilt and trolley wire replaced the remaining third rail sections in the Scranton yards. These improvements cost about $30,000, but as the extent of the Depression deepened such programs were sharply curtailed.

In 1931, less than $1,000 was expended, mostly for electric traffic signals at two crossings in Pittston and Wilkes-Barre. These devices eliminated the need for crossing watchmen and by the end of the decade they had been installed at most of the remaining grade crossings. Rarely used crossovers, such as those at Broad Street and Butlers, north of Pittston, and at Hilldale, were removed.

The second track from Connell Junction to Maple Street in Scranton on the over-the-hill route was retired in 1934 and the remaining section between that point and Laurel Junction was removed in 1939. Except for the Dunmore local car, which ran only to Maple Street, this section of the line was now rarely used. Four of the old frame houses in Scranton, originally rented to employees, were demolished in 1936 as they had become rundown and no tenants could be found. The brick structure in Wilkes-Barre, originally used as an express depot and more recently as a warehouse, was sold to Swift and Company with the proceeds turned over directly to the Guarantee Trust Company.

Dunmore can most simply be defined as a rectangle cut out of the northeast corner of Scranton. Although the Laurel Line served Dunmore, it reached only the southeastern edge of the more populated portion of the borough. Scranton Transit Co. streetcars serviced the heart of Dunmore, such as flamboyantly attired 402 (ABOVE) loading inbound at Dunmore Corners in 1939. *Michael J. Lavelle, Edward S. Miller Collection* **September of the same year saw L&WV's car 19 (RIGHT) on the Dunmore run descending the steep grade just below Laurel Junction. In this view from Scranton's Harrison Avenue bridge, the DL&W railroad main line appears to the left.**

George Krambles

In keeping with its reduced prospects, the company gradually trimmed its roster during the Depression years. In December 1930, the line listed 28 passenger cars, six motorized freight and work cars, and 44 other units of rolling stock. With the reduction in schedules, only nine cars were needed for base service and 12 for peak periods. During a representative week in March 1933, most of the type M cars were out of service and the bulk of the schedules were handled by the 31-series steel cars, which were averaging about 800 miles. Among the combines only one type M car, number 111, saw any use, although the three steel combines averaged almost 1,400 miles in that week. The two remaining express cars, 202 and 204, ran slightly over 100 miles during that seven-day span. All three locomotives were active with 403 evidently being a favorite. It ran every day and chalked up 525 miles.

With a surplus of equipment on hand, the L&WV gradually retired older cars. In 1931, coaches 12 and 13 and combo 112 were removed from the roster. Four years later, three more of the venerable type M's were withdrawn as was flatcar 302. The motors and electrical equipment from number 16 were retained as spares, while the body was used to store materials at the powerhouse until November 1939 when it was burned. Motors and electrical equipment from coaches 17 and 18 were transferred to locomotive 402, which finally was relieved of its obsolete type 50F motors and was now on a par with 401 and 403, which had been upgraded earlier. The body of car 17 was sold for $100, while 18 remained on the property until it was burned late in 1939.

As of 1938, coach number 10 and baggage trailer 101 were retired. The next year, coach 21 was retired and its body sold for $100. At the end of the decade, the Laurel Line owned 20 passenger cars, two express cars, three locomotives, 38 freight cars and four nonpowered work cars. In a general cleanup of the property, the hulks of numbers 10 and 104 were burned at the end of 1939. A side benefit of the reduced roster was the slight saving realized in rental fees for Ohmer fare registers. In 1932, the company maintained 22 of these devices, but by 1940 only 17 were required and two of these were spares.

The cost of maintaining the equipment remained steady throughout the decade. The 15 active passenger cars collectively required an average of about $32,000 in materials per year to keep them up. Wages for the shop force averaged a very reasonable $21,000. One practice which the company, unfortunately, did not change was its habitual tardiness in submitting reports to various agencies. Officials at the Pennsylvania Department of Internal Affairs were particularly disturbed each year when they had to pry the required reports from Murphy's office.

For instance, in May 1930 the department sent several letters, politely requesting a prompt return of the report. These were ignored until the chief of the agency resorted to a telegram and threatened a stiff penalty. Even though Murphy was away that day, an unsigned report was promptly mailed and a signed one was submitted on the president's return. The same process had to be repeated each year so that in 1933 the exasperated public official told Murphy, "Your firm is one of a very few in the state which has not yet filed a report." It seemed only the threat of the $200 fine could motivate Murphy to complete his paperwork.

As slow as Murphy was in his business dealings, the Interstate Commerce Commission (ICC) was even worse. J.A. Chisholm, assistant treasurer, received a letter dated November 21, 1939, from Frank Fowler, director of the Bureau of Accidents at the federal agency. Fowler said the ICC had learned that Acting Vice President W.E. Higgins, who was listed in its files as the man in charge of destroying records, was no longer with the firm. He asked who had replaced him. Chisholm dryly replied the next day, "Mr. W.E. Higgins left the service of this company in 1915," and added the auditor took care of the records. No doubt management took some comfort from the knowledge that they were not alone in their tendency to be tardy.

The L&WV Power Company Declines. The electrical system gradually deteriorated as accidents, reduced maintenance and time took their toll. A severe storm in early July 1933 flooded the area around the Plains substation, forcing the company to shut off all power from that facility. The Avoca substation had to carry the extra load, but the voltage dropped so sharply on the southern end of the line that trains could not maintain their schedules and all runs suffered delays until the flood waters subsided. Some Lehigh Valley trains, which connected with the Laurel Line cars, were held so passengers could make the transfer. In the same month, the 75-kilowatt transformer at Virginia substation burned out. Because of the reduced industrial demand, it was not replaced. Although the power system was rated at 650 volts DC, the actual average voltage on line varied from 550 to 570 volts.

Imitating its parent, the power company also lost money in the decade. A 1937 report noted the company served 22 customers, had 28 employees, ran two steam turbines fueled with 22,393 tons of anthracite and lost $1,260. To cover itself in the event of an accident, the company carried a $50,000 per accident insurance policy at a premium of $7,898.

Problems with Freight. The Depression had less of an effect on freight traffic than passenger revenue, but problems abounded nonetheless. Package express gradually declined despite active solicitation, advertising and new overnight less-than-carload (LCL) service from Philadelphia to Wilkes-Barre, operated by the Lehigh Valley Railroad. The basic reason was the reduced business activity in the region so that merchants simply had less need for the service. Growing truck competition also contributed to this general downturn. In one week in mid-July 1933, eight firms and individuals applied for permission to operate trucks between Scranton and Wilkes-Barre, which the L&WV strongly opposed.

Annual revenue fell from $5,730 in 1930 to a low of $3,838 in 1935, recovering slightly to $4,378 in 1940. Mail continued to add slightly to the company's depleted coffers. An average of about $2,200 per year was earned, although it is questionable if any profit was accrued. For instance, in April 1930, on many days the company carried only three to seven pouches of mail for which it received slightly under 19¢ per trip. Even when as many as 30 pieces were carried, only 38¢ per trip was earned.

The company served an impressive list of industries, although some of them were not major shippers. Among the more important were: the Socony-Vacuum Oil Company, Jacob Dold Packing Company, Spruks Brothers Lumber Company and Frank D. Fields Company, all in Scranton; Moses Griffith Lumber Company in Plains; and Wilson and

The buildings of central Scranton form a backdrop as 401 and eight cars head south at the entrance to the tunnel on May 25, 1939. The power house is on the left, next to Roaring Brook. Part of the car house can be seen through the trees to its right. On the far left is the right-of-way of the Dunmore branch descending the hill. Although the L&WV station cannot be seen, the Lackawanna Station adjacent to it appears just over the Spruce Street Bridge. *Edward S. Miller*

Company, Swift and Company and the Lehigh Beef Company in Wilkes-Barre. Coal was received over the line by the L&WV power company and fuel oil by the Pennsylvania Power and Light Company in Wilkes-Barre. The carrier maintained 15 hopper cars for interchange service and 16 additional cars for on-line service only. Interchange agreements were finally reached in 1932 with the New York Central and Pennsylvania Railroads for their routes east of Buffalo. Interchange tracks were located in Scranton with the DL&W, in Wilkes-Barre with the Lehigh Valley and in Pittston and Dunmore with the Erie. In 1931, the L&WV received 1,139 cars from the DL&W, 1,010 from the Lehigh Valley and 12 from the Erie. Outbound carloads, consisting mostly of coal, were distributed as follows: 589 to the DL&W, 156 to the Lehigh Valley and 327 to the Erie. These ratios remained roughly the same throughout the decade.

Although many commodities were carried, coal was by far the most important. The major mining firm was the W. Moffat Coal Company, which had a breaker at Virginia. Several other firms produced coal but on a more sporadic basis. The number of cars handled varied widely depending on the market and labor conditions in the coal fields. In

Patrick Murphy rejected the idea of substituting motor buses for the likes of handsome interurban car 33, here crossing the Roaring Brook trestle outbound from Scranton in September, 1939. The line managed to keep its equipment presentable despite half-full cars such as this.
George Krambles

Clattering through the switch leading to the freight station, a two-car northbound train accelerates out of the Pittston station in 1940. Crowning the hill in the right background is Pittston High School, and below it is the Pittston Stove Works. As in other locations, the Laurel Line was the newcomer and could not attract freight here. The Erie Railroad, not the L&WV, served the stove works. Today most of the factory has been dismantled, but the William Street side remains as a retail store. Until April, 1924, trolleys of the Scranton Railway ran on Broad Street between the L&WV and the stove works. *Edward S. Miller*

An elderly lady is more interested in the photographer than the passage of northbound car 35 as it travels on the canal right-of-way at River Street station and approaches the Susquehanna River in September, 1939. Indicative of the decline of the railroads is the placement of a passenger shelter on one of the Lehigh Valley Railroad's freight tracks. The manicured lawn on the right stretches to the Luzerne County Court House.
George Krambles

1934, 4,434 cars were hauled; in 1936, 8,038; in 1937, 9,919; but only 1,478 in 1938; and 1,826 in 1939. The failure of the Moffat Coal Company in the fall of 1937 accounted for the sharp drop.

A dispute with the DL&W arose over per diem charges in 1936. The previous winter the DL&W had shipped a number of cars to the Moffat breaker in Virginia. Because of the severe weather, the coal was frozen and could be removed only by burning ties and automobile tires beneath the cars, a process which took some time. As a result, considerable per diem charges were built up. During the same period, the DL&W held three L&WV hopper cars in its yard at Hoboken for a total of 1,520 per diem days. The L&WV wished to offset the $2,452 it owed in charges to the DL&W by deducting the amount owed to it. The basic quarrel centered around the L&WV's contention that run-of-mine coal shipped inbound to a breaker was exempt from per diem in that, once the coal was processed, it would be sent back as an outbound load. The Erie Railroad accepted this arrangement, but the DL&W did not. Furthermore, the electric line charged only six cents per gross ton while the steam road received from six to 19 cents per gross ton, depending upon the distance covered, and they could haul the coal in long trains, whereas the L&WV averaged only three cars. It appears that the L&WV was forced to accept the DL&W position.

Various attempts were made to stimulate business by encouraging coal companies to ship their products from washeries and breakers located on the L&WV. Essentially, however, the depressed state of the coal industry resulted in only minimal success. Compounding the Laurel Line's problem was the fact that it was basically a terminal switching operation with virtually no bridge traffic. Consequently, it received only a minimal return, despite the large number of cars it handled. Revenues increased gradually from $101,101 in 1931 to $192,049 in 1937. The failure of the Moffat Coal Company in 1938 reduced revenues to $92,473, although they recovered to $130,514 in 1940.

The Bus Line Weathers the Depression. Although the coal traffic, by its nature, was basically confined to the rails during this period, the L&WV was vulnerable to competition from trucks and buses for most of its other services. A report made to the federal coordinator of transportation in 1933 analyzed the competitive position of the company. At that point, the only scheduled motor bus service paralleling the rail line was the route between Scranton and Pittston operated by the L&WV itself. Murphy estimated this line diverted earnings of about $5,000 yearly from the interurban cars. Another $5,000 was lost to various charter buses which carried groups at low rates unmatched by the electric carrier.

More competition was offered by a number of interstate bus lines, operating to such points as Buffalo, New York City, Philadelphia, Washington and Pittsburgh. These lines maintained agencies in the small towns along the railroad and paid commissions to the shopkeepers for tickets they sold. Murphy assumed that people riding these buses would otherwise have taken the Laurel Line to either Scranton or Wilkes-Barre and steam trains beyond. He estimated the L&WV lost about $50,000 from this diversion.

Perhaps more serious was freight which was lost to motor trucks. These vehicles reached all points served by the electric road and, of course, could travel far beyond its limits. They tended to siphon off the more profitable tonnage from larger shippers, leaving a comparatively small amount for railcars. Furthermore, their rates were considerably lower than the rail lines. Some of the truckers not only delivered the goods to the customer's door, but even carried them up two or three floors or down to the basement without additional charge.

Because of the Depression, a growing number of owner-operated trucks appeared, and their rates not only undercut the railroad but the common carrier truckers as well. The L&WV reduced its LCL rates on many occasions in an effort to regain lost traffic, but in each instance the truckers made greater reductions. As a result, LCL tonnage declined from 11,646 in 1924 to 1,494 in 1932, and revenue in the same period fell from $34,484 to $8,682.

Murphy rejected the idea of substituting motor buses for the electric cars. Although buses could reach some stations easily, in other places people would have to walk from one-half to one mile to reach the highway. The buses were slower and less comfortable than the high-speed railcars and would require higher commutation fares. The Laurel Line's own bus route between Scranton and Pittston took 45 minutes to cover the 11 miles, while the railcars negotiated the same distance in only 21 minutes. Murphy admitted that motor trucks could accommodate the local LCL freight, but the rail line would have to be maintained in any event for carload freight. He therefore concluded it would be uneconomical and unwise to substitute buses and trucks for rail service.

Other problems perceived by the president of the L&WV were competition from private automobiles, general unemployment, low divisions of through freight rates, high per diem charges and the erratic performance of the anthracite coal mines, which shut down frequently due to strikes and the loss of markets to new fuels. "Thousands of people depending upon the mines for their livelihood are idle and as the L&WV railroad operates through the heart of the anthracite region the effect upon its earnings are ruinous," Murphy concluded.

As solutions to these unsolvable problems, he proposed strict enforcement of state laws prohibiting "thumbers" or hitchhikers on the highways. He also suggested the steam railroads should be more liberal with their rate divisions and relieve shortlines of per diem charges. He advocated federal control of the mining industry, including authority to prevent strikes, which he felt would result in a substantial reduction in the price of anthracite and thereby stimulate business recovery. As a final point, Murphy wanted electric railways to be eligible to secure loans from the Reconstruction Finance Corporation.

Murphy's analysis of the problems besetting the L&WV was essentially sound and it illustrated the magnitude of the carrier's difficulties. His proposed solutions, however, were politically naive and economically unfeasible. While all railroads faced similar difficulties to some degree, the situation was compounded by the Laurel Line's short length, heavy dependence on the passenger trade, lack of through-freight traffic and basically poor financial condition.

Partly as a means of maintaining its relative position, the L&WV continued to develop its highway services during the 1930s. The name of the subsidiary was changed late in 1932 from the Laurel Line Bus Company to the Laurel Line Transportation Company. Murphy continued as president; G.W. Esslinger as vice president; S.C. Stivers, secretary; and

J.A. Chisholm, treasurer. Among the directors were William von Puhl and S.C. Stivers of New York; Murphy of Scranton; and Sproul and W.H. Lippincott of Philadelphia. In April 1930, the bus line was extended from Moosic to Scranton; two years later, the company applied for an amended charter, which would permit it to carry freight and merchandise. A certificate of public convenience was issued on July 31, 1933, and approved by Governor Gifford Pinchot three weeks later.

The four Mack buses, numbers 15, 17, 18 and 19, which were acquired in 1926, were retired in 1932 and 1933. They were replaced by number 20; a Mack 1930 model BC, seating 29, at a net cost of $6,950; and number 21, a White 1930 model 65 vehicle, with 25 seats. In 1932, three additional Whites were purchased, numbers 22 and 23, also model 65, with 29 seats, and number 24, a model 613 with 21 seats. Two years later, Dodge model K34 vehicles with 19 seats were placed in service and numbered 25 and 26. Three additional Dodge machines were added in 1936. They were model LF37A with 21 seats.

Evidently, vehicle life was short for in 1937 four new Ford model 70, 27-seat buses were purchased for $4,500 each from the Transit Bus Corporation of Philadelphia. Seven older vehicles were retired, two Macks, four Whites and a Dodge, which the company hoped to sell for $4,500, matching the purchase price of the new buses. Among the retired buses were 20, 22, 23 and 26. They were sold to Mendenhall Bus Service of Glendora, New Jersey. In 1937, a model 81B Ford, seating 27 and costing $4,744, replaced a 1936 Dodge. Another Dodge was sold to a Nanticoke firm in 1938 and an additional Ford was acquired in December 1939. It had 27 seats and cost $4,744.

To handle the freight service, the company maintained two 1½-ton trucks. As with the buses, these vehicles were frequently replaced with newer models. Dodges were preferred. The snowplow, which was purchased in 1926 for $250 for attachment to a truck in order to clear the roadway from the Avoca garage to the main highway, was rarely used. It was sold 10 years later for $100. Unlike the buses, which at first penetrated no further south than Pittston, the motor trucks operated a pickup and delivery service between Scranton and Wilkes-Barre.

On the average, five buses were required for regular daily schedules. Usually, at any one time, the company had a fleet of seven on hand. Occasional special service was offered, but the line made no serious effort to exploit the charter market. In 1935, the company won a contract from the Treasury Department to carry Works Progress Administration workers from Dupont and Pittston to Highway Route 665, where a WPA project was underway. The work continued through the spring of 1936 and the company realized a payment of $6,527 for its services. In 1930, the nine workers employed by the transportation company were paid an average of $37.80 per week. Under the pressure of the declining economy, this fell to $36 by 1932 and to $28.50 two years later. There was no union. The average number of employees during the decade was 18.

Financially, the Laurel Line Transportation Company was marginal at best. At the beginning of the decade, the company earned $58,426, which barely covered expenses. In the succeeding years, income averaged about $51,000 from passenger operations. Motor freight service, which began in 1934, contributed an average of almost $8,000 annually. Profit margins were slim and in some years losses were recorded. The greatest profit was realized in 1931 when $14,757 was gained, but the next year the company showed a loss of $1,082. In 1940, a profit of $1,915 was made. The transportation company showed the same tendency to be late in meeting financial obligations as its parent. In June 1938, the city clerk of Pittston warned the line that all franchise privileges granted to the firm would be discontinued if its 1937 and 1938 license fees were not promptly paid. The old saying, "Like father, like son," aptly applied to the relationship between the railroad firm and its corporate stepchild.

LAUREL LINE TRANSPORTATION COMPANY
INCOME STATISTICS

YEAR	BUS PASSENGER REVENUE	OTHER PASSENGER REVENUE	FREIGHT REVENUE	TOTAL REVENUE	OPERATING EXPENSES	NET INCOME
1930	$57,311	$ 839		$58,426	$48,045	$10,758
1931	66,849	1,328		68,242		14,757
1932	53,506	1,486		54,993		(1,082)
1933	49,462	2,304	$ 370	52,136		(4,261)
1934	54,304	2,834	5,856	62,994		1,921
1935	52,249	5,259	6,711	64,219	65,322	(1,102)
1936	51,807	9,963	7,529	69,823		2,123
1937	49,054	3,186	8,863	61,236	55,418	7,288
1938	50,303	2,531	9,073	61,906	58,414	3,492
1939	49,213	2,109	9,294	60,616	53,575	6,890
1940				57,288	55,413	1,915

Source: Laurel Line Transportation Co., Operating Revenues from June 30, 1924; Income Statement 1938-1947, L&WV Collection.

Scranton's Spruce Street Bridge was always an ideal vantage point from which to observe the comings and goings of the Laurel Line. Looking west (ABOVE) number 36 can be observed leaving the terminal for Wilkes-Barre while the Dunmore branch car enters the loop. The interchange traffic for the Lackawanna was so heavy on this particular day that it choked all but one track in the yard. The L&WV yard is now the site of the University of Scranton's athletic field. Looking east (RIGHT) the Dunmore branch car can be seen as it reaches the bottom of the hill. One car provided service every half hour while the other surviving wooden car was held as a spare. Double track remained at this point to provide access to the freight yard. Both pictures were taken in the summer of 1940. *BOTH: Edward S. Miller*

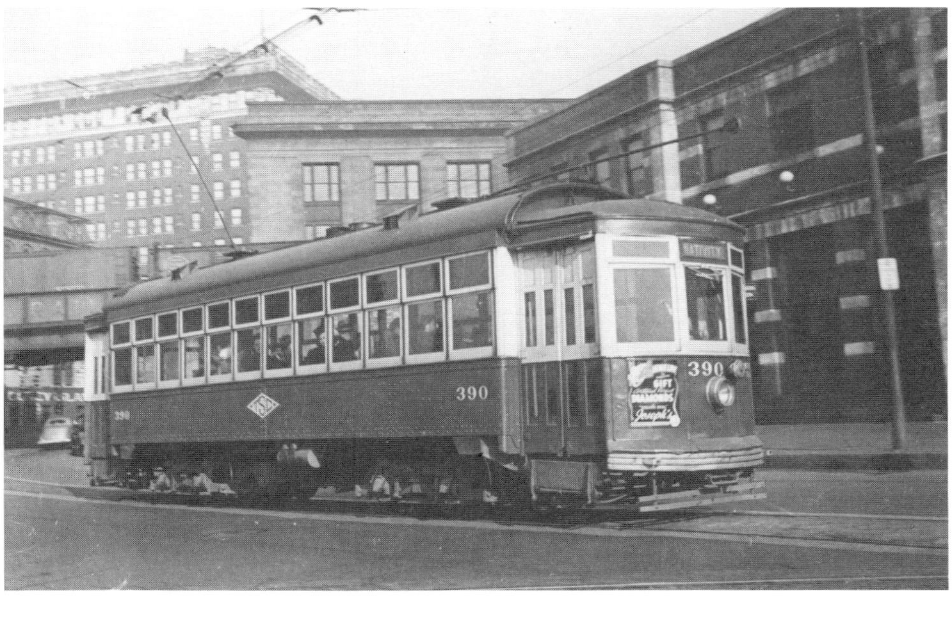

Passengers could avail themselves of connecting service offered by Scranton Transit Company's Nativity line to reach the L&WV station, immediately to the right of the trolley. Interestingly, the south end of this line ended on Beech Street at Crown Avenue, directly over the Laurel Line's tunnel. Until 1924, Scranton Railway cars to Pittston, which competed with the Laurel Line for local traffic, rolled down this same street. *Stephen D. Maguire*

The stockholder who asked "Is the road running?" obviously had not seen this two-car train heading under the Erie Railroad underpass just south of the North Pittston station. The Independence Day holiday in 1940 required the additional equipment to service Rocky Glen Park. *Edward S. Miller*

Heavy Losses Strain the Laurel Line. While the Laurel Line Transportation Company was able to report slight operating profits, it was not saddled with the crushing debt of the railroad corporation. The ravages of the Depression increased this burden to such an extent that abandonment became a distinct possibility by the end of the decade.

Despite the stringent operating economies described earlier, the financial condition of the carrier worsened as the Depression deepened. As shown in the chart nearby, the line reported a deficit in every year following 1929. Freight held up better than passenger revenue, with $101,102 realized in 1931 and $130,514 in 1940, with sharp variations in particular years caused by changes in the mining industry. The operating ratio climbed from 81.6 percent in 1931 to an alarming 100.8 percent in 1940. During the same period deficits ranged from $63,561 to $67,114. Even more disturbing was the change on the corporate balance sheet from a surplus of $32,136 in 1931 to a deficit of $521,688 in 1940.

Management reacted to the financial crisis in several ways. The Hoover Administration had retreated from the concept of "rugged individualism" to the extent that the president approved congressional legislation creating the Reconstruction Finance Corporation on January 22, 1932. This agency was a sort of government "superbank," which was empowered to lend money to hard-pressed insurance companies, banks, railroads and other enterprises. The New Deal administration of Franklin D. Roosevelt continued and expanded this program in an effort to reverse the spiral of business failures.

Faced with growing losses, Murphy decided to apply for an RFC loan in December 1934. However, there was little chance that the federal agency, which was engaged in lending money for such monumental projects as the San Francisco Bay Bridge, the Los Angeles water supply system, and the electrification of the main line of the Pennsylvania Railroad, would approve funds for the lowly L&WV, which lacked any political clout. The electric line was forced to get by as best it could.

Under the circumstances, it was inevitable that the holders of the company's securities should suffer. With the withdrawal of Ford, Bacon, and Davis and the deaths of other principals in the 1913 reorganization, Murphy, who was elected president in 1930, took over direction of the company. Among the board of directors during the period were William von Puhl, John T. Collins, Jr., J.A. Emery, E. Clarence Miller, Senior and Junior, Sproul and S.C. Stivers. There were about 20,000 shares of preferred and 48,000 shares of common stock outstanding, with the largest blocks held by George Bacon, George Davis, Walter Lippincott, Miller, Sproul and Murphy. The rest was scattered among various individuals who held relatively small numbers of shares.

Stockholder discontent at the absence of dividends was largely ignored by the harried management. Even during the prosperous twenties, there were occasional letters from curious stockholders requesting financial information and reports on the company's condition. For the most part, they received no reply, nor did the line distribute an annual report. By the 1930s, when dividends ceased, more stockholders wrote, often on the bottom of their proxy forms. One complained in 1937, "Has this stock any value? I can find no quotation and they tell me that it has no value. How can I find out?" Another asked, "Is the road running? Is it showing a profit? Is it improving? Do you expect to resume dividends?" In

1938, one irate stockholder wrote, "You are a hard crowd to get information from. Are you drawing large salaries, say $1,500 a year?"

One persistent individual who owned 50 shares of preferred stock and who had written every year since the mid-1930s demanding information was last heard from in 1940, when he penned, "This road is such a miserable failure that there is no value in this stock, although preferred."

So many people neglected to send proxies, either through anger or frustration, that the company had to make special efforts to secure the minimum number for a quorum. The only response management made to critical stockholders was to send them the current income statement and balance sheet, which revealed all too plainly the mounting deficits.

While the stockholders could safely be ignored, the bondholders were another matter. If interest on their securities was unpaid or reduced, they could force the line into bankruptcy and possible liquidation. Ownership of the bonds was scattered among a large group of people with only the banks and a few brokerage firms holding significant amounts. Fortunately for the L&WV, the bondholders were willing to forego part of their interest, possibly because adding another company to the long list of bankrupts would accomplish little and might even be deemed unpatriotic by New Deal politicians.

Beginning in August 1933 the company asked the bondholders to allow a partial interest payment on their semiannual coupons. Usually this involved canceling $10 or $15 of the $25 interest payment due. From 1933 through 1940, the company canceled a total of $225 in interest per bond. On the average, about 98 percent of the bondholders accepted the reduced payments. Murphy blamed the declining anthracite business on which the company so heavily depended, as well as the Depression, for these measures. Even so, L&WV bonds were almost worthless on the financial market. In December 1940, one of the first mortgage 5 percent bonds with a par value of $1,000 sold for only $11.

Another step taken by management to ease the financial crunch was a policy of purchasing its 5 percent first mortgage bonds. As of 1938, the company had bought bonds worth $400,000 in principal at a cost of only $162,584. The aim was to reduce indebtedness in the hope that the declining earnings could cover the interest payments on the outstanding bonds. Unfortunately, this was not possible. During the decade of the 1930s an average yearly default of $130,000 was reported because of the inability to meet interest payments.

The company's bookkeeping methods occasionally caused criticism. During most of the decade, the L&WV power company reported slight profits as did the Laurel Line Transportation Company. The Guarantee Trust Company, holder of most of the first mortgage bonds, demanded to know in 1938 where the dividends from these firms had gone, saying they should have been sent to the trustee. Somewhat wearily, the secretary explained that the dividends were only a bookkeeping device and since these companies were wholly owned by the L&WV, the slight profits of its subsidiaries were used to reduce the deficit of the parent. As hard-pressed as the line was, it was forced to pay higher taxes, since local municipal governments and agencies increased rates to meet the challenge of the Depression. Local taxes jumped from $4,500 in 1932 to $5,300 in 1935 to $10,700 in 1938.

By 1939, the L&WV was nearing the end of its financial resources. While freight traffic held promise, passenger service appeared to be close to extinction. The outbreak of war in Europe in September 1939 would dramatically change the situation. As the United States prepared for possible conflict, the Depression came to an end, business revived, and patronage on the red electric trains began to increase. From its narrow point of view, the future looked brighter for the Laurel Line.

Railway Income Statement

	1931	1932	1933	1934	1935	1936	1937	1938	1939	1940
REVENUE										
Passenger	410,592	296,243	243,897	252,219	228,090	243,112	230,834	215,978	202,969	190,252
Freight	101,102	110,582	119,611	148,165	141,563	166,839	192,049	92,473	104,957	130,514
Other	12,573	10,729	10,806	10,270	13,844	17,894	21,995	9,868	9,069	7,732
Total	524,268	417,555	374,314	410,653	383,499	427,844	444,878	318,319	316,965	328,498
EXPENSES										
Maintenance of Way & Structures	51,827	46,082	41,012	52,808	53,932	63,381	66,224	32,045	37,910	43,356
Maintenance of Equipment	57,236	50,997	49,662	52,019	50,473	51,287	55,179	49,809	51,157	54,489
Power	114,928	108,259	97,782	95,501	103,369	113,995	94,468	102,697	91,833	78,905
Transportation	169,055	147,883	144,544	146,766	143,962	149,224	146,228	120,899	118,098	121,541
Other	53,112	48,646	50,464	43,225	39,834	44,119	46,164	40,067	41,074	40,326
Total	446,158	401,868	373,465	390,320	391,570	422,007	408,263	345,517	341,072	338,517
NET REVENUE	100,524	33,654	12,935	34,012	4,424	17,589	37,585	(−15,882)	(−13,320)	(−2,810)
Taxes	4,634	4,506	4,807	4,471	5,391	13,923	10,673	15,797	15,916	16,372
GROSS INCOME	146,849	99,523	56,780	81,407	66,609	74,743	90,989	38,429	32,366	44,371
Interest	220,234	219,175	219,120	214,151	212,064	208,000	137,052	112,183	113,296	108,839
NET INCOME	(−82,255)	(−126,592)	(−167,114)	(−137,236)	(−154,205)	(−141,464)	(−63,561)	(−82,724)	(−86,763)	(−70,189)
DIVIDENDS PAID	0	0	0	0	0	0	0	0	0	0
CORPORATE SURPLUS OR DEFICIT	32,136	(−91,812)	(−222,134)	(−215,574)	(−311,374)	(−317,101)	(−316,560)	(−355,901)	(−458,089)	(−521,688)

Source: Financial Report, Before the National Railway Labor Panel Emergency Board, 1925-1942, Miller Collection.

Wartime sinkings of ships in the North Atlantic totally disrupted the normal ocean shipping of petroleum to eastern markets. Consequently, railroad tank cars were used to carry the trade. The mainline railroads ran solid trains of oil from the southwest. The Laurel Line did its share for the war effort. Illustrated are seven of these now virtually forgotten vehicles being switched prior to leaving the Scranton yard. *George Arents Research Library, Syracuse University*

The Laurel Line Goes to War. World War II was a mixed blessing for the L&WV. Restricted by gasoline rationing and rubber shortages, the traveling public returned in large numbers to the electric railway. In 1940, the road carried 1,036,010 passengers; five years later, 2,125,701 people rode the line, an increase of 105.5 percent. Passenger revenue jumped from $190,252 to $443,502, an increase of 133 percent.

Despite this dramatic increase in patronage, only minor schedule changes were made. The September 15, 1940, time card was superseded by a new one on December 8, 1944. The 30-minute headways, which had been in effect from 6:00 A.M. to 7:00 P.M. were extended one hour, with 60-minute service from 8:00 P.M. until midnight. Sunday headways were reduced from hourly to 30 minutes between noon and 10:00 P.M. Fares remained unchanged. Local travel predominated; only 3 percent of the riders purchased interline tickets.

A considerable number of men entering the military forces used the Laurel Line, since all draftees began their careers in Wilkes-Barre with a pre-induction physical and subsequent swearing-in ceremony. Men from Lackawanna County and parts of Luzerne, Susquehanna and Wyoming Counties utilized the line in both instances. In August 1942, Murphy reported the road was carrying about 1,000 of these individuals per month. For the entire wartime period, General Passenger Agent W.L. Cairns estimated the line hauled more than 50,000 such passengers. In contrast to World War I, the L&WV labor force itself was barely affected by the draft, since most of the men were now beyond military age.

Also receiving a reprieve from the war was Rocky Glen Park. Since it was the only such amusement center in Lackawanna County, it attracted weary workers who could not use their cars to reach other diversions. The Laurel Line's revenue from ticket sales to the park jumped from $10,000 in 1940 to $16,000 in 1941 and $28,000 in 1942. By 1945, the line received $38,143 from the 136,414 fares it carried.

The divided ownership of Rocky Glen and the L&WV's prior involvement with it caused some uneasy relationships. Partly because of the split management and also because of self-interest, the Laurel Line undertook to book picnics, place advertising and generate publicity for the park. W.L. Cairns, the general passenger agent of the line, was responsible for this, but Benjamin Sterling, Jr., managing director of half of the park, objected to the arrangement and tried to persuade Murphy to end it. The Laurel Line president refused, and told Sterling in March 1941 that the L&WV's aid and publicity greatly benefited the park. Since the other owners backed Murphy, Sterling was forced to drop the matter.

Murphy had his own grievances against Sterling. The park leased a section of land from the railroad and was sometimes late in making its annual rental payments. Murphy wanted the park to maintain a fence next to the right-of-way to prevent children and others from wandering onto the tracks, but wartime shortages and lack of attention resulted in large holes and consequent trespassing. After several incidents where cars nearly struck individuals, Murphy demanded repairs be made and the area policed to prevent vandals from inflicting damage. Thus, the L&WV had an ambivalent attitude toward this attraction along its line. Probably Murphy's earlier experience in managing the park and the financial losses he sustained as a result embittered him toward the operation and colored his views toward it.

Freight operations were not stimulated by the war. Revenue fluctuated between $105,467 in 1940 and $93,179 in 1945. Carloads of merchandise handled during the same period varied from 2,490 to 2,367. Coal carloads ranged between 3,165 and 1,053. Major shippers and receivers of carload freight included: Scranton Coal and Ice Company;

the Lehigh Beef Company; Swift and Company; Thomas C. Thomas, a wholesale produce dealer; and Wilson and Company. Coal remained the major commodity hauled by the L&WV. Some of it went off-line to the Erie and Lehigh Valley connections but a considerable amount was burned in the L&WV powerhouse. LCL traffic remained relatively minor with about 1,200 tons shipped to local industries and about the same amount forwarded to interline connections.

The trucks of the Laurel Line Transportation Company handled about 1,300 tons of freight annually. Murphy noted the war did not really increase the L&WV's freight traffic, since no new industries located along the line and existing industries were not greatly affected by the war. Highway competition still remained, despite the Laurel Line running a pickup and delivery service for local freight customers.

The population of the anthracite region was declining, which adversely affected freight earnings, since major shippers, such as the meat packers, no longer had as large a demand for their products. Oil traffic was lost to trucks because of a tank car shortage. All freight handled originated or terminated on the L&WV; there was no bridge traffic. Although freight would ultimately be the line's sole source of revenue, the L&WV remained a terminal switching operation.

Surprisingly, there was no drastic increase in the number of employees, despite the growth of passenger traffic. The basic reason was that the cars had carried only light loads during the Depression and the additional wartime patrons were easily absorbed. Office, maintenance, shop and transportation forces grew slightly, but in 1944 the line rostered only 132 employees versus 127 in 1940.

Labor relations remained peaceful during the war. Murphy pleased the workers in June 1941 by restoring half of the 10 percent wage cut made in 1932. He was able to do this because the low price of coal reduced the cost of electricity. Under terms of an agreement with the Brotherhood of Railroad Trainmen in December 1943, the line paid motormen and conductors 89¼¢ per hour, with the usual 4¢ differential for service on electric engines. Trainmen on the one-man Dunmore cars received 91¾¢ per hour, and brakemen earned 83¼¢ per hour. The railroad payroll totaled $253,782 in 1945. The line drew a reprimand from the United States Department of Labor in 1944 for failing to pay powerhouse workers time and a half for overtime.

Wartime Shortages Hinder Improvements. The increased revenue from wartime business gave the L&WV an opportunity to catch up on much-needed maintenance, which had been cut to the bone during the lean Depression years. Ironically, however, shortages of material and labor permitted only minimal improvements.

New 85-pound rail was installed along 3,500 feet of the northbound main line, below the Dupont station, in 1940. Shorter sections of the southbound track in the same vicinity were replaced, and some new rail was also spiked down along the main line near the Scranton terminal. The abandonment of the Miners Mills streetcar line along Pennsylvania Avenue on Saturday, October 19, 1940, by the Wilkes-Barre Railway permitted the removal of the crossing frogs there.

The heavily used trackage in the tunnel also needed replacement, and Murphy placed an order for new rail in 1942. Steel was naturally much in demand by then, and the Office of Defense Transportation (ODT) gave the line a low priority. Some two years went by, and still the federal agency would not approve the order. Finally, in the fall of 1944, after two separate instances of broken rails in the tunnel, Murphy warned of a possible disaster. Local ODT officials telephoned Washington and obtained the necessary authorization. About 139 tons of new rail were ordered from the Carnegie rolling mills of United States Steel Co. and installed in January 1945.

The deteriorating trackage forced numerous slow orders at various points along the line, in addition to the permanent ones at Laurel Junction, the Pittston curves and the Prospect Viaduct. Usually limits of 20 or 25 mph were placed, although a speed of only 5 mph was ordered at the Plains station partly because children crossed the tracks here on their way to and from school. In another safety move, a derail was installed on the over-the-hill line track, about 150 feet north of Connell Junction. This section was used for occasional storage of freight cars and was not normally in service.

In order for a newer railroad to obtain permission to cross an older, established one, the newer carrier generally assumed all responsibility for maintenance of the crossing. Hence we witness a Laurel Line crew, using crane 802 and locomotive 402, replacing the worn crossing diamonds where a previously established spur from the Lehigh Valley Railroad's Canal Branch was crossed by the Laurel Line outside the Miner Hillard Milling Co. plant near N. Washington Street in Wilkes-Barre. The Lehigh Valley Railroad, which had a higher war materials priority than the L&WV, was able to obtain the diamonds when the L&WV could not. The work was done and paid for by the L&WV in 1945. *Edward S. Miller*

The slipshod mining methods practiced by the Rhonda Coal Company caused serious disruptions at the powerhouse in August 1944. The mining company discharged its refuse into Roaring Brook, and the resulting silt clogged the pumps and condensors to such an extent that freight service had to be suspended and passenger trains were delayed. The L&WV had drained the pool behind its supply dam in June, but it was filled with silt in only two months. When the company ignored Murphy's plea that it clean up its operations, the Laurel Line resorted to a civil suit to prevent further dumping. Partly as a result of the service interruptions, the Pennsylvania Coal Company canceled its power contract with the L&WV, effective at the end of the year. Coal, as usual, was a mixed blessing to the Laurel Line, providing freight revenue on one hand but sometimes adversely affecting service.

The shabby condition of the cars and stations attracted the attention of Philadelphia attorney E. McLain Watters, who was preparing for the inevitable reorganization of the line. He urged Murphy to clean and paint the structures to give a good appearance to the property. The president replied that he normally had a crew of three painters, who worked on the cars during the winter and on the buildings during the warmer months. Labor shortages reduced this force to only one man and a helper, who soon fell behind. Like so many other firms, the L&WV had to make do as best it could during the critical war years.

The Dunmore Branch Is Abandoned. Despite the surge of wartime traffic, it was clear that a decline could be expected once hostilities ceased and restrictions on driving were removed. The short Dunmore line, with its odd, three-terminal configuration, had always been marginal at best. With another financial reorganization in the offing when the bonds of the company matured, abandonment of this stub operation became attractive to the Laurel Line's managers, especially Watters, who was beginning to play an active role in the line's affairs.

One problem which helped to bring the matter to a head was increasing public criticism of the danger from the exposed third rail, which was responsible for a number of fatalities from careless or unwary trespassers. The original fences which had lined the right-of-way had long since fallen into disrepair. Another factor which contributed to the hazard was that the railroad was no longer a central part of the public's concern. People were increasingly oriented toward automobiles and highways, and what had been regarded in the past as a technological and engineering marvel which served the general public was now seen by many as an obsolete and dangerous mode of transport which threatened the welfare of some members of the community, especially children and incompetents. The Laurel Line's minimal efforts at public relations and the obvious aging of its plant and facilities contributed to this image.

Passenger loads were still high on Memorial Day, 1945, as numbers 116 and 31 rolled out of the south portal of the Scranton Tunnel. Although the war with Germany had recently ended, bringing the photographer home from Europe, war restrictions continued for several more months.
Edward S. Miller

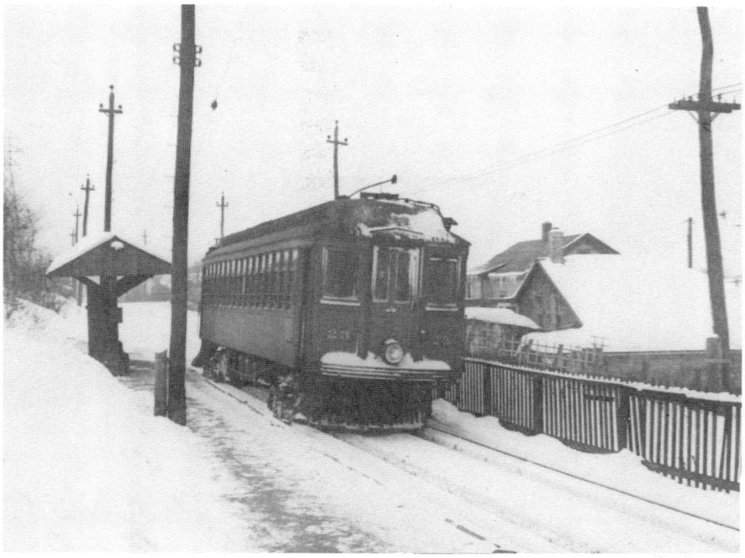

The south end of the Dunmore Branch was this unpretentious shelter adjacent to the former northbound track at Maple Street. This stop had been designated South Scranton until 1914 when all mainline trains were routed through the tunnel and a new South Scranton station was erected on that route. No longer needed, the former southbound track lay undisturbed for many years under a mat of brush, being dismantled gradually as the railroad needed the rail or splice bars elsewhere. The decaying fence separates the two tracks and prevented patrons from an unpleasant encounter with the third rail. The track being used was intact for the remaining 3400 feet to Connell Junction and could be used in emergencies to bypass the tunnel until 1945. In fact, it was still intact as late as 1947 to Maple Street, and about 500 feet remained north of Connell Junction in 1952. Today a foot bridge over Interstate 81 marks this site.

LEFT: Stephen D. Maguire
BELOW: Edward S. Miller

In the summer of 1944, a series of accidents involving contact with the third rail prompted the Pennsylvania Public Utility Commission (PUC) to pressure the line to take steps to reduce the hazard. PUC engineers inspected the property and insisted the line institute a program of fencing or covering the rail. Murphy estimated it would cost about $100,000 to make necessary changes. He wrote to the Philadelphia Transportation Company and the Pennsylvania-Reading Seashore Lines, both of which operated third rail lines, for information about their systems. The latter corporation estimated it would cost the L&WV about $250,000 to reequip its cars and plant in the same manner as the Shore Fast Line route. Murphy told Watters that fencing was unobtainable until after the end of the war but that abandonment of the Dunmore branch and the remainder of the over-the-hill line would eliminate about 4½ miles of the most troublesome trackage and reduce state and public criticism.

Apparently for the first time, Murphy ordered an analysis of the Dunmore route, which involved 4.36 miles of single track. Service was offered from 5:44 A.M. to 12:15 A.M. on weekdays, with 30- and 60-minute headways, and from 6:25 A.M. to 11:30 P.M. on Sundays and holidays, with hourly headways. The fare was 10¢ cash, or 8¢ with strip tickets; there were no transfers. For the years 1941 through 1944, about 92,000 passengers were carried annually, paying an average of $8,800 yearly. Freight added about $1,100 per year. Operating expenses ranged from $19,000 to $26,000, and annual losses of $10,194 to $15,631 were reported, with little prospect for improvement when the war ended.

After a conference among Watters, Secretary S.C. Stivers and Murphy in Scranton in September, it was decided to apply to the PUC for permission to abandon the line. This decision was formally ratified by the executive committee of the L&WV on October 5. However, nothing was ever accomplished simply on the Laurel Line. The effort to abandon the unprofitable branch turned into an embarrassing comedy of errors before the rails could be removed.

As a preliminary step, Murphy contacted the Erie Railroad, which readily agreed that it could route empty coal cars to the electric road over the interchange near Pittston just as easily as it could using the Dunmore interchange. Next, Murphy met with Harry H. Dartt, president of the Scranton Transit Company, who also registered approval and who promised to extend his Dunmore and South Side bus routes to serve the vicinity of the stations at Ash, Myrtle and Maple Streets. In fact, however, the transit company would drag its corporate feet and its failure to readily extend its bus routes would impede the L&WV efforts to abandon. The formal petition was filed with the PUC in Harrisburg on November 6, and hearings were scheduled in due time.

A Dunmore Pictorial

The Dunmore Branch was originally that trackage from Laurel Junction, east of Scranton on the "Over-the-Hill" line, northward to the Erie interchange in Dunmore. In the wake of all of the runaways in 1914, the railroad revised schedules so that all mainline trains went through the tunnel. Thereafter, schedules showing the Dunmore Branch included the southern stub of the "Over-the-Hill" line. After leaving Scranton, trains ran up the hill to Laurel Junction and then turned north onto the old Dunmore Branch. At Dunmore, the motorman walked to the opposite end of the car with his control handles and then ran the car south to Maple Street, bypassing Laurel Junction in the process. Again the car was reversed in order to return to Laurel Junction and Scranton. During the period of one-man operation, all station platforms were on the east side of the track. Car 19 (LEFT) enters the original Dunmore Branch at Laurel Junction in 1945. Compare this view with the one on page 30 showing the line when it opened and with the illustration on page 186 during the construction of the expressway network. The same car is depicted in 1939 southbound, 500 feet above Latches on the single track connection between the old Dunmore Branch and the old mainline. The track to the left was once the double track mainline to Laurel Junction, over which the car will return to Scranton. The photographer was standing at the Moosic Street bridge, close to three present Interstate 81 interchanges. The new Sheraton Motor Inn is located about two blocks behind the photographer.

Although almost entirely within the City of Scranton, the north end of the branch ran through a wilderness with only limited residential development to generate riders. Near the end of its life in 1945, the weeds grew unchecked between the rails as car 19 (ABOVE) passed the deserted Nay Aug Park station. Although Scranton Transit ran frequent and full streetcars to the other side of the park, the Laurel Line handled very few riders. An average of ten people per trip at this time was normal. The next stop to the north was at Myrtle Street (RIGHT), where two women wait next to the ramshackle shelter for a southbound car. The old Luna Park station was within a "stone's throw" of Myrtle Street. Today Interstate highway 81 either parallels or uses the Dunmore right-of-way. *BOTH PAGES: Edward S. Miller*

Running about ten minutes late, the 1:30 p.m. car from Scranton flies around the curve between Moosic and Avoca stations on August 10, 1945. The original Moosic station at Powder Mill Rd., illustrated on page 22, was located here. *Edward S. Miller*

While these moves were being taken, Murphy was chagrined to receive a letter from the Guarantee Trust Company of New York, the trustee and holder of the company's first mortgage bonds and debentures, on which interest was much in arrears. Purely by accident, someone in the office sent a copy of the executive committee minutes of October 5, referring to the proposed abandonment plea, to the trust company. The bankers were quick to point out that all proceeds from salvageable material, including that reused elsewhere on the L&WV, must be promptly sent to the trust company under the terms of the 1913 mortgage.

Murphy sheepishly explained to Stivers that ". . . we sent the minutes . . . by mistake" and asked for his advice. The secretary remarked that had the minutes not been sent ". . . we might have got away with the amount received from the sale of scrap. It looks now as though we would have to deposit with them any moneys received from the sale of property from the Dunmore branch." Thus, Murphy's hopes of using the salvageable material elsewhere on the line without contacting the trustee were dashed.

Meanwhile, the PUC slowly conducted hearings. The city of Scranton protested the abandonment because of uncertainty about the willingness of the transit company to provide substitute service. The bus company cited rubber and gasoline shortages as justification for its delays. Finally, on September 24, 1945, the PUC gave its blessing and the last runs were scheduled for Thursday, October 25. On that date at 12:15 A.M., car 19, which for 10 years had shared its burden with car 23, made the last trip under motorman Thomas Shreck's control. As the coach neared the Scranton terminal, the rowdy crowd proceeded to strip it of curtains, flags, seat covers and other items as souvenirs. The coaches were stripped of usable parts, and for the first time no type M cars remained on the active roster.

Just as Murphy was making plans for the last runs, he was jolted by a letter from the Interstate Commerce Commission in Washington asking why he had not filed a petition with the federal agency seeking its approval for the abandonment. The Laurel Line president responded that he did not think it was necessary since the line in question was purely local in character. Attorney Watters, however, said he ". . . felt sure it was necessary to advise the ICC," but he had not done so earlier since Murphy's office felt it was unnecessary. He now prepared to go to Washington to represent the L&WV at the inevitable federal hearings.

As if this were not enough trouble for the harried L&WV president, he now learned that an on-line shipper, the Scranton Redi-Mix Concrete Company, had filed a belated protest with the PUC against the abandonment of the line. The concrete company provided minimal business for the Laurel Line; in 1944 it received 14 cars, and thus far in 1945 it had unloaded only 16 cars of cement. The plant was located almost two miles up the hill from the terminal, near the Waldorf Station, and unless its objections could be satisfied, the railway would have to maintain the most difficult and dangerous section of the line it wanted to abandon. Nonetheless, the PUC scheduled a conference before it would permit abandonment.

The meeting was duly held at the end of November and involved representatives of the DL&W and Erie Railroads as

A two-car train consisting of numbers 115 and 34 heads toward Wilkes-Barre at the South Pittston station near S. Main and Plank Streets. The short branch of the Erie Railroad leading into Pittston is on the hillside to the left of the train. The string of coal-laden hopper cars in the background is on the interchange track between the L&WV and the Erie. All of the railrod tracks in this view have been dismantled in the intervening four decades. *Edward S. Miller*

well, since it was hoped they might offer substitute service. Various options were discussed, such as constructing a siding from the nearby Erie branch, or having the DL&W service the plant with one of its diesel locomotives, or even moving the factory to another location. No conclusion was reached, and the PUC advised the L&WV to continue running freight trains to the plant as needed until further notice. The publicity generated by this affair prompted a number of former commuters to demand the restoration of passenger trains as the substitute bus service offered by the Scranton Transit Company was deemed inadequate.

While Murphy fumed, attorney Watters was making progress with the ICC in Washington. The federal agency insisted that its cumbersome rules be followed, which included the posting of public notices and a hearing, which was held in February 1946. Finally, on March 29, the ICC granted approval of the petition to abandon the line. The cement company was forced to make other arrangements and track removal got underway on May 5.

Some of the rail would be used on the Moosic freight branch, which was being constructed about that time. At last the L&WV was free of its costly line to Dunmore, and the last remnants of the old hope of extensions to the north and profitable bridge traffic disappeared with it. In due course, the Central Scranton Expressway and interstate highway 81 would obliterate most traces of the electric road in the area.

The Bus Line Struggles On. The war produced operating pressures on the bus subsidiary. Shortages of gasoline, tires and other items plus increased passenger loads plagued the company. At the beginning of the conflict, the line had six 27-seat buses, numbered 30 through 35, all built by Ford between 1937 and 1940. Three 1½-ton trucks, numbered 3,

4 and 5, completed the roster. Most of the buses were used on the 10-mile Pittston–Scranton run, although a few trips were made between Dupont and Wilkes-Barre, a distance of eight miles. Traffic was very light and this route was discontinued in 1943 in order to concentrate on the Scranton line.

A few buses continued to run between Scranton and Wilkes-Barre, mostly to prevent competition. Although schedules were constantly adjusted and every effort made to reduce mileage, the overtaxed fleet of aging buses frequently broke down. Only one vehicle was available as a spare, and not until 1945 did the Office of Defense Transportation permit the line to add another Ford bus to its roster.

Financially, the transportation company reported a profit during the period. Revenue rose from $61,250 in 1941 to $106,400 in 1945, and net operating income jumped from $8,200 to $32,500 during the same time. Like its rail counterpart, however, the bus company would face greatly reduced prospects when automobiles once more became available at the end of the war.

Financial Problems Mount. Despite the wartime surge in traffic, the financial condition of the L&WV remained tenuous, although few realized how quickly the line would decline once peacetime conditions returned. As shown in the chart, business peaked in 1946, as returning servicemen, orders for goods and materials long unavailable and the continuing shortage of automobiles benefited the electric road. But the basic fact remained that the line was not earning enough to pay its funded debt interest charges, even with the increased moneys generated by wartime demands.

Only in 1944 and 1945 did black ink appear on the net income line in the company's ledgers, and these small amounts were woefully insufficient to cover the huge amount of deferred interest claims. Even more alarming were the rapidly approaching dates of February 1, 1948, when almost $1 million in 6 percent debenture bonds matured, and August 1, 1951, when nearly $2 million in first mortgage bonds were due. The coffers were bare, and with the almost certain likelihood that traffic would decline in the postwar period, the L&WV faced a grim future.

In addition to worrying about a future reorganization, the line's officers managed to get involved in a long-running dispute with the Commonwealth of Pennsylvania over a tax claim for $2,500. The money was due as a result of the L&WV increasing its capital stock in 1922 and thereby becoming liable for a so-called bonus payment to the state. Rather than pay this claim and be done with it, the management decided, in typical fashion, to fight it on technical grounds. Eventually the state prevailed, and in 1944 the company had to remit not only the principal but the interest which had accrued over the years.

But changes were in the offing for the L&WV. Longtime Treasurer J.A. Chisholm resigned in June 1944 to accept another job in New York. Murphy, now 75 years of age, was obviously nearing the end of his career. The 250 stockholders were quiet, but they had become immune to the absence of dividends and the fact that no market existed for their stock. The bondholders were still accepting some partial interest payments because they had no choice, and their bonds could be sold only at discounts of 60 percent or more on the over-the-counter market.

The fate of the company hinged largely upon the goodwill of the bankers and trustees, who held many of the securities, and the skill of the new managers who were waiting to take over. Of these, the most significant was the Philadelphia attorney, E. McLain Watters, chairman of the executive committee. The postwar years would be crucial for the Laurel Line and would determine how much longer the red electric cars would roll through the mine-scarred valleys of northeastern Pennsylvania.

Railway Income Statement

	1941	1942	1943	1944	1945	1946
REVENUE						
Passenger	200,620	305,455	435,258	410,736	443,502	468,386
Freight	137,235		127,847	141,359	119,116	144,811
Other	7,379		7,142	7,001	7,717	8,235
Total	345,234	460,169	507,247	559,096	570,335	621,432
EXPENSES						
Maintenance of Way & Structure	44,180	47,327	84,631	63,233	62,159	71,279
Maintenance of Equipment	52,077	51,980	62,350	65,393	68,537	74,620
Power	76,838		106,179	94,505	97,476	107,210
Transportation	122,978		168,691	190,102	188,617	224,303
Other	42,825		45,265	45,466	55,386	51,238
Total	338,898	386,671	467,116	458,699	472,175	528,650
NET REVENUE	12,947	73,498	109,593	108,026	107,921	105,672
Taxes	16,887	20,380	26,968	27,056	24,487	28,931
GROSS INCOME	60,364	117,043	111,445	121,682	158,426	111,047
Interest	120,934	138,945	114,397	58,807	91,711	164,052
NET INCOME	(−68,264)	(−21,902)	(−2,952)	62,875	66,715	(−53,005)
CORPORATE SURPLUS OR DEFICIT	(−565,261)	(−587,163)	(−489,826)	(−426,970)	(−112,814)	(−267,738)

Source: Financial Report before the National Railway Panel Emergency Board, 1925-1942; Miller Collection; L&WV Railroad, Comparative Income Account, 1941-1950; Breakdown of Income Account, 1943-1946, L&WV Collection.

Home on leave from the army, photographer Edward S. Miller spent August 11, 1945, hiking along the Laurel Line. Shortly after ten o'clock that morning he snapped a southbound coach (RIGHT) at Connell Junction. The single track remnant of the original route, probably last used extensively during the tunnel relining in 1917, veers to the right behind the car. In this 1945 view, all traces of the former rock crushers at this site are obscured by brush. By early afternoon, the photographer waited in this idyllic setting (ABOVE) north of Rocky Glen Park, its serenity disturbed only by the periodic chirping of cicadae, the creaking of the expanding rails, and the pungent odor of hot creosote from the ties, until this two-car train clattered around the curves. The previous evening's newspapers carried front page headlines announcing that the second of two awesome atomic bombs had been unleashed on the Land of the Rising Sun. Four days later, on August 15, 1945, Japan surrendered unconditionally and all war labor restrictions were immediately lifted. The following day, gasoline rationing ended and Americans resumed their love affair with the automobile, forcing the Laurel Line to struggle as never before for its survival.

The sunshine sparkles on the newly fallen snow as Combine 114 heads north at Mill Creek at the Wilkes-Barre city line during February, 1946. To the left of both the Susquehanna River and the Laurel Line tracks is the Lehigh Valley Railroad's Canal Branch, built on the right-of-way of the former North Branch Canal, and the narrow gauge rails of the Lehigh Valley Coal Co. *Edward S. Miller*

6

The Postwar Years: Decline and Bankruptcy

THE POSTWAR WORLD brought rapid changes in international affairs, lifestyles and modes of transportation. Following the death of Franklin D. Roosevelt on April 12, 1945, Harry S Truman of Missouri assumed the office of President. Feisty, aggressive and determined, he battled not only his traditional opponents, the Republicans, but many of his fellow Democrats as well. Elected in his own right in a stunning upset in 1948, Truman declined to run again in 1952. In that year, for the first time since 1928, a Republican, Dwight D. Eisenhower, won the presidency, and the Grand Old Party assumed power in Washington the following year.

Americans were increasingly preoccupied with foreign affairs, as wartime cooperation between the Western powers and the Soviet Union gave way to mutual suspicion and mistrust. The expansion of communism into Eastern and Central Europe, and soon China, provoked a firm reaction from the United States. Soon the Cold War was underway, marked by a race for supremacy in atomic weapons, the Berlin Blockade and in June 1950 the Korean War, a three-year struggle for domination of that remote peninsula, which eventually ended as a stalemate. Since the administration insisted the Korean conflict was a police action, it was fought in limited fashion, and thereby had little effect on companies such as the Laurel Line.

At home, the postwar era was marked by a relentless monetary inflation, an exodus from the cities to suburbia and a growing flood of automobiles on the nation's highways. The age of the automobile, and its allies, the bus and truck, had truly arrived.

The triumph of the motor car doomed the nation's few remaining interurbans, whose ranks had already been decimated by the Depression. In the decade or so following the end of the war, most of the remaining lines were either abandoned or converted to diesel freight service. Included among them were such once prominent carriers as the Shore Fast Line in New Jersey, the Baltimore and Annapolis in Maryland, the Piedmont and Northern in the Carolinas, the North Shore Line and the Chicago, Aurora and Elgin in Illinois, the Utah lines, the Pacific Electric in California, and the West Penn Railways and the Lehigh Valley Transit in Pennsylvania.

The position of the Laurel Line was especially difficult. Never strong financially, unable to realize its promoters' dreams of serving as a bridge line for freight, heavily dependent on passenger revenues and burdened with a huge debt, it was located in a region which was undergoing an economic transformation and whose population was declining. Anthracite coal had been the mainstay of this part of the Keystone state for more than a century, but coal was "king" no longer, as its markets slowly collapsed.

The decline had been underway for some years, but it was exacerbated by frequent strikes, rising prices and the availability of relatively cheap oil and natural gas. Anthracite had been used more for commercial and home heating in the northeast than by industry or power plants; it was especially vulnerable to competition from fuel oil and gas. Homeowners and businesses alike welcomed the opportunity to dispense with shoveling or stoking bulky coal and removing ashes, when heat could be regulated by simply turning a thermostat.

In 1924, Lackawanna and Luzerne counties produced 55,103,550 tons of anthracite; by 1930 this had fallen to 42,768,955 tons. The decline continued during the Depression decade and by 1940 the two counties marketed 31,931,479 tons. Production stabilized during the war, but after the conflict, a rapid drop in anthracite mining took place. In 1950, the two counties sent 23,639,223 tons to market; by 1952, the last year of L&WV passenger service, the tonnage had fallen to 19,078,497.

The drop in production was matched by a decline in employment in the region's coal fields. In 1924, the two counties had 99,154 miners. Six years later, more than 97,000 were still at work, but by 1940 only slightly more than 60,000 were employed, a decline of 38 percent. In 1950, there were only 44,180 workers in the industry, and two years later this had fallen to 38,229, a drop of 61 percent since 1924.

The erosion of the anthracite market had drastic effects on the population of Lackawanna and Luzerne counties. Peak population figures were recorded in 1930, when Lackawanna had 310,397 people and Luzerne 445,109. On the eve of World War II, in 1940, the counties numbered 301,243 and 441,518 respectively. By 1950, with mining collapsing and

Mine products comprised a major share of the L&WV's freight traffic. Typical was this view of a train of mine waste, or culm, at South Scranton awaiting clearance from the dispatcher to enter the single track tunnel into Scranton proper. Most seats in the two-car southbound passenger train were filled with fun-seekers headed for Rocky Glen Park on this summer Sunday afternoon in 1948. *Edward S. Miller*

more lucrative jobs elsewhere, Lackawanna's populace had fallen to 257,396 and Luzerne's to 392,241. The next census in 1960 revealed population figures of only 234,531 and 346,972 respectively. In 30 years the two counties had suffered a loss of 23 percent in population.

The same trend was true of the two leading cities of the region. Scranton, which had 143,433 persons in 1930 recorded only 125,536 20 years later, and by 1960 the city had but 111,443 inhabitants. Wilkes-Barre in 1930 numbered 86,626; this fell to 76,826 in 1950 and to 63,551 by 1960. Over the span of 30 years, Scranton lost 22 percent of its population and Wilkes-Barre lost 27 percent. Many workers, especially young people, left the region to seek opportunities in more promising localities. Public officials and businessmen tried, with limited success, to lure new industries to the area, but the streets of Scranton, Pittston and Wilkes-Barre assumed a "frozen-in-time" aspect, with few new buildings or developments which were transforming so many other urban areas. This economic stagnation, coupled with increasing ownership of automobiles, improved roads, and changing lifestyles stimulated by such things as television and shopping centers, could only bode ill for the struggling Lackawanna and Wyoming Valley electric line.

A New Management for the Laurel Line. The first year after World War II, 1946, was a good one for the L&WV. Returning servicemen, the pent-up demand for goods which had been unavailable during the war, the still limited availability of automobiles, and a spirit of optimism following the successful conclusion of the great struggle against the Axis powers all contributed to make a banner year for the Laurel Line. A total of 2,260,864 passengers were hauled, the largest number since 1929. Although operating revenues exceeded $634,000, a deficit of $53,000 was recorded, largely because of interest on the funded debt. Despite this, Murphy felt it was time to step down. He had been employed by the L&WV for 31 years and, having guided the line through Depression and war, he was ready for retirement.

Methodical as he was, Murphy had planned for this change for some time. The man he had groomed as his successor, Watters, was elected to the presidency in December 1946 and assumed responsibility for the electric line on the first day of 1947. Born in 1881 in Akron, Ohio, to a family of Welch descent, Watters was graduated from the law school of the University of Pennsylvania in 1904. Although admitted to the bar, he became active in the investment banking business, where he made a reputation by organizing and supervising bondholders' organizations, especially of commercial, public utility and electric railway corporations.

In 1942, Watters returned to an active legal practice and began to take an interest in the affairs of the L&WV as well. He purchased stock and was elected to the board of directors, where he became very friendly with Murphy, who persuaded him to succeed him in the presidency. At the same time, Watters was also chairman of the board and vice president of the *Altoona and Logan Valley Electric Railway* (Penna.) and a voting trustee of a minority stock group of the *Fort Dodge, Des Moines and Southern Railway* (Iowa), as well as serving in various capacities on several utilities. Despite his busy schedule, Watters agreed to accept the challenge of the L&WV presidency, partly because he knew a reorganization would soon be necessary and partly because he had a genuine interest in the line and wanted it to continue in operation, preferably with passenger service.

Watters resumed the longtime L&WV tradition of absentee management, since he continued to maintain his law office in Philadelphia, although he traveled frequently to Scranton to oversee the affairs of the Laurel Line. An affable, friendly man, with a good sense of humor and an air of polite gentility, he was about six feet tall, with dark hair and a spare frame. His son, E. McLain Watters, Jr., was elected secretary.

Watters was fortunate to have capable men assist him. The turnover among the operating employees was very low, and most of the officers had considerable experience as well. The on-site general manager was William L. Cairns, who had been with the company almost from its inception, serving as general freight and passenger agent for many years. A native of Massachusetts, he lived most of his life in nearby Clarks Summit.

Another longtime employee was John J. Hennigan, who had risen through the ranks from office boy, paymaster, claim agent and chief clerk to superintendent of transportation in 1943. A native of Scranton, he died in 1961. Other regulars were Treasurer J.J. Loftus, Chief Engineer George D. Shrader, and Master Mechanic D.C. Shellabarger. None of the officers received excessive salaries; in 1949, except for Watters who earned $5,900, they all received less than $5,000 each. Varying percentages of each man's salary were charged to the three L&WV companies, the railroad, power or transportation company, depending upon his responsibilities.

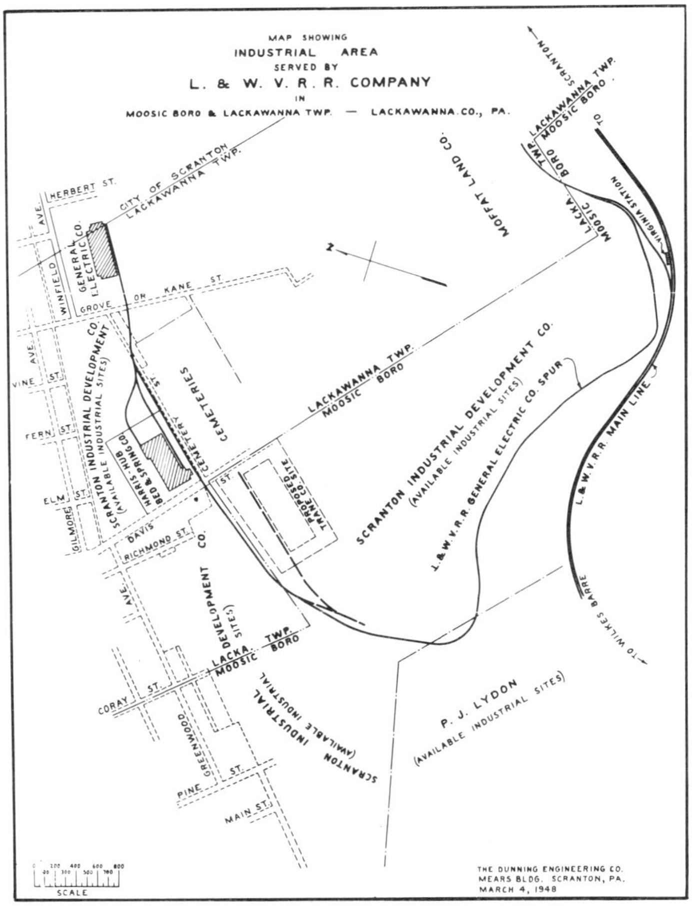

George Arents Research Library, Syracuse University

The Last Extension.

One reason for Watters' optimism was the development of a 350-acre industrial park in Lackawanna Township, immediately south of Scranton, which was served exclusively by the L&WV. Aware of their overdependence on the vulnerable anthracite trade, civic leaders and businessmen in Scranton had formed several organizations to lure new enterprises to the region. Among them were the Scranton Industrial Development Company and the Scranton Lackawanna Industrial Building Company, both actively backed by the Chamber of Commerce.

In 1944, the Chamber asked the L&WV to estimate the cost of building a branch to serve the industrial area. The new extension would leave the main line at Virginia station and use the old right-of-way to the former Moffat Coal Company breaker, and then continue about 3,000 feet beyond, for a total distance of about 2½ miles. Because of the stiff grades, the branch would be entered through a switchback arrangement at Virginia, whereby trains would back up a long stub track before proceeding up the twisting right-of-way through the borough of Moosic to the industrial park. Murphy estimated it would cost about $25,378 for the work, using mostly rail salvaged from the soon-to-be abandoned Dunmore branch.

No action was taken immediately, but by 1946 the development organizations had succeeded in persuading the General Electric Company to build a plant on the site. Permission was granted by the Public Utilities Commission to cross the three streets in the vicinity at grade. By this time, the Dunmore line had been abandoned, so rail and ties from that route were laid on the new spur. About 75 percent of the rail was of 75-pound weight; the remainder was 90-pound. Additional used ties were ordered from a dealer, but when they arrived, the L&WV judged them to be inferior, good for only about five years. Reverting to its traditional practices, the L&WV tried to get the dealer to give them the ties free, with the company paying the $1,282 freight bill. It is not known how the matter was resolved, but the ties were used.

Although not quite finished, the spur was opened early in 1947, using overhead trolley wire for power. Soon men unloading boxcars at the General Electric siding complained they were receiving electrical shocks. It was determined the problem was the absence of rail bonding at the splices. Not until mid-February was the bonding completed, and not until the end of September were crossings bucks erected at the three street intersections.

The Minooka Branch

The Minooka Branch was opened in 1946 to serve the General Electric plant and other factories in the newly-developed industrial park. The shift to motor transportation rendered this newest part of the Laurel Line obsolete by 1983. Engine 403 (ABOVE) is shown moving two revenue cars on the new line west of the old Moffat Breaker site in July, 1947. The rear pair of cars contain ballast to support rails on the expanding line. Locomotive 402 (BELOW) is shown shifting cars at the GE plant.
BOTH: Edward S. Miller

No. 402 (RIGHT) waits with two cars at the foot of the steep grade where the Minooka Branch left the mainline at Virginia station. One brakeman is on the ground setting a derail to prevent any runaways from fouling the mainline. The four wheel tower car had been used so extensively on this line recently that the company left it to the right of the train. In a few days it would find use again by crews stringing trolley wire over the spur into the Trane plant. Passenger trains were never scheduled over the Minooka Branch but several chartered railway enthusiast excursions did tour the industrial complex. Posed beside car 32 (ABOVE) in 1949 is one such group. Many of the patrons lived in the area. Included from left to right are: Michael Golden, motorman (in vestibule); Jerome J. Burke, conductor (on ground); Bud Chase; Harold Geissenheimer; John Endler; James Langan; John Gabriel; J.J. Hennigan, L&WV superintendent; William F. Ellis; and Edward S. Miller. *BOTH: Edward S. Miller*

General Electric, which made dishwashers, was soon joined by two other firms. The Harris Hub Bed and Spring Company opened a plant in 1947 for the construction of beds, rails, cabinets, wardrobes and wire crating, and the following year the Trane Company erected a factory to build convector radiators. All these firms received inbound carloads of steel and other materials and shipped carloads of their finished products.

On rare occasions, a passenger train negotiated the stiff grade of the branch. For example, in March 1948, a "prosperity special" was run to show prospective buyers available sites. But mostly the line was kept busy with freight. By 1951, General Electric was providing about 1,500 cars annually, Harris Hub was using about 450 cars, and the Trane Company provided the Laurel Line with about 212 cars.

Other shippers provided sufficient business to keep the freight crews fairly busy in 1950. The Poloron Company, which made army kits and fire bombs, built a plant on the branch which required about 250 cars a year. Firms on the Scranton end of the line included the Standard Beef Company, which received about 150 cars; Fiegleman Waste Materials Company, Kurtonick and Shaffer and Albert Nivert and Company, which shipped or received about 260 cars of scrap iron and steel, paper and rags; L&WV Power Company; Keystone Cinder Company and other small organizations, which collectively sent about 2,000 cars to cinder block manufacturers in other cities in the northeast. In Wilkes-Barre, the three on-line customers gave the company about 675 cars. They were the meat packers, Swift and Company and Wilson and Company, and Thomas C. Thomas, a wholesaler of fresh fruits and vegetables.

One negative feature of the Wilkes-Barre business was that often the refrigerator cars of perishables arrived on the DL&W interchange track in Scranton in lots of only one or two cars. They had to be moved immediately, which involved calling extra crews. The revenue received did not offset the cost of the special handling involved.

At the same time, anthracite shipments were declining sharply. In 1947, the road handled 4,250 cars; this fell to 2,309 cars in 1948, 1,598 cars in 1949, and only 426 cars in 1950. Clearly, the future of the freight business would depend on other commodities than the traditional anthracite.

Roadway Projects Keep Work Gangs Busy. While work was underway on the Minooka branch, other projects had to wait for attention. The line had become shabby and rundown after 15 years of neglect. Watters was aware of the importance of safe operation and a neat appearance, but the fiscal restraints of the company limited the extent of any improvements.

The usual mine cave-ins vexed the roadway gangs, with particularly severe ones near North Pittston and Midvale in 1947. Slow orders were issued whenever these incidents occurred, until the roadbed could be filled. Also troublesome were the street crossings in Wilkes-Barre, which needed repaving and repairs, provoking complaints from residents, the city and the state highway department. One of the worst was at North Franklin Street, where one homeowner reported cracks and damage to his property caused by "excessive vibration" when cars rattled over the worn and loose rails. Cairns could only reply that, when time permitted, work crews would rebuild the crossing. Also in need of replacement was the trolley wire in Wilkes-Barre; new wire was finally purchased in 1947.

Even more serious was the condition of the bridges. As early as 1944, the Erie Railroad had complained repeatedly that a bridge on the Number 6 Washery branch was badly in need of repairs. Other bridges needed new ties and paint, including those at Avoca, Prospect and Mill Creek. The painting at Avoca was awarded to an outside contractor who bid almost $6,800 for the job. The other projects were handled by the two work gangs maintained by the line, one operating from Wilkes-Barre and the other from Scranton. They were normally kept busy with routine maintenance, such as replacing ties, adding tie plates, replacing switches and frogs, removing snow and all the other chores required on any railway.

Locomotive 403 leads a typical nine-car freight train north past Virginia Station in June, 1947. The switch to the right leads to the newly-opened Minooka Branch, which would ultimately be one of the last active portions of the railroad. *Edward S. Miller*

The second longest bridge on the Laurel Line measured 554 feet in length (one foot shorter than the height of the Washington Monument) and rose 30 feet above the ground. The Prospect Viaduct, located in Plains Township, 1,200 feet north of the Wilkes-Barre city limit, spanned the Lehigh Valley Railroad's Canal Branch and its passenger mainline, the narrow-gauge tracks of the Lehigh Valley Coal Co., freight tracks of the Central Railroad of New Jersey, River Street and a Wilkes-Barre Railway Co. trolley line. Its erection in 1903 by the McClintic-Marshall Construction Co. required relocation of the trolley line and the highway. The bridge is shown (ABOVE) supporting the weight of a single car in 1950. The photographer was looking south from a position on the Lehigh Valley tracks. The other view (RIGHT), taken in 1940, looks north from the city line along River Street and shows a trolley on the Hudson line and the Prospect Colliery to the right. Prospect Viaduct appears in the background. Trolleys competed with Laurel Line trains from 1903 to 1941.

BOTH: Edward S. Miller

Despite their efforts, they were never able to catch up. In mid-1951, Roadmaster Sheridan warned that the Prospect viaduct near Wilkes-Barre might have to be removed from service because ". . . the ties are rotted and worn so bad that the rail is two inches away from the face of the ties." Trains had been theoretically limited to five miles per hour for a period of 10 years on this span.

Just as serious as bridge maintenance and a source of rising criticism from the public was the danger from the uncovered third rail, aggravated by the poor condition of the roadway fences, which required constant attention. The wooden posts were old and rotting, the wire was rusty, and vandals were continually cutting and breaking it. The greatest concern expressed by residents and civic leaders was for the safety of children. One irate parent in Avoca warned the Laurel Line its unguarded third rail was an example of "criminal negligence" and that he would ". . . prosecute to the fullest extent of the law" should any harm befall his three sons.

Similar letters were received from people in all the on-line towns. The problem seemed insoluble, because other residents would cut the barriers in order to avoid walking to road crossings or bridges. General Manager Cairns pleaded for help from the Avoca chief of police after newly repaired fences had been destroyed. Only the abandonment of electric power in 1953 provided a real solution.

Despite the hazards, the line had no serious accidents in its final years. Minor incidents did occur, as when two cars were struck by lightning in Wilkes-Barre yard in July 1948, ". . . on account of the trolley poles being up." As a result, orders were given that poles were to be hooked down on out-of-service cars. Heavy rain caused a retaining wall to collapse near the Spruce Street bridge in Scranton in May 1946, derailing a northbound train. None of the 13 passengers was injured, but the DL&W wrecking crew had to be summoned to rerail the car.

Six weeks later, an inattentive motorist drove in front of car 114 at the West North Street crossing in Wilkes-Barre. The heavy interurban pushed the car 40 feet down the track and wrecked it, but fortunately the occupants sustained only minor injuries. In September 1951, a southbound car carrying nine people was stalled for more than three hours in the tunnel when the third rail overturned. Susquehanna river flooding occasionally disrupted service. Buses were substituted during all of these mishaps. Undoubtedly, the experience of the crew members, most of whom had years of service on the carrier, contributed to the good safety record.

The Search for Funding Continues. Increasing reliance on carload freight and declining business made some structures and property superfluous. In 1946, the Great Atlantic and Pacific Tea Company leased an area in the terminal for $1,500 a year; in 1949, the Standard Beef Company assumed the lease for the same rental. The Scranton freight house burned on December 17, 1947, causing more than $15,000 in damages, most of which was covered by insurance. The structure was repaired and divided; part was leased to the Ideal Confectionery Company and the remainder was rented to Dennis Pappas of Union City, New Jersey, as a warehouse ". . . for the storing, ripening, receipt, sale and distribution of bananas and other produce" for $600 annually. For a while, the line received a bonus from this arrangement, since the fruit was shipped in railroad refrigerator cars and unloaded at the site.

The financial pinch forced the line to seek funds from any available source. In 1950, the remainder of the over-the-hill line between Scranton yard and Connell Junction, and unused yard trackage, was sold for scrap. The $3,400 realized was immediately applied to the overdue tax bill. A parking lot was constructed on part of the terminal property in Scranton and leased for $125 monthly. The newsstand in the station realized $300 annually in rental payments. All this helped, but it was not enough to gain solvency.

The same could be said about the mail contract. The line continued to carry closed pouches of mail either on its regular passenger trains in coaches or combination cars, or in the express cars. Eleven trips were scheduled on weekdays and two trips on Sundays, with extra trips when necessary. The post office paid only four to six cents per mile on each trip, so total revenues amounted to only about $3,000 a year. For example, train number 2, between Scranton and Wilkes-

According to the photographer's notebook, it was 3:35 p.m. on Sunday, June 20, 1948, when this two-car train rolled past Connell Junction. Having topped the 1.4 percent grade south of the tunnel, this train is on the longest level stretch on the entire railroad — one-half mile in length. *Edward S. Miller*

In a scene rendered desolate by mine refuse cast off from the abandoned Pennsylvania Coal Co. Butler Breaker, last operated by the Volpe Coal Co. under lease, express car 202 rumbles southward parallel to U.S. Route 11 north of Pittston during the summer of 1947. The crude bridge over the track was used by trucks entering the breaker, out of sight to the right. When the Laurel Line opened in 1903 it competed with Scranton Railway Co. trolley cars which ran beside the highway to Pittston. In the almost four decades since this vignette was exposed, highway traffic was partly diverted to the Pittston Bypass, to the right of the picture. The culm bank, on which the photographer stood, was leveled. The rails were removed. Finally, the Pittston Plaza shopping center was erected on this site.

Edward S. Miller

Barre, made 313 one-way trips per year, amounting to a total of 5,934 miles. The mail rate was five cents per mile, which meant the L&WV realized $296.72 annually for carrying mail on this train. Nonetheless, since the trains were running in any event, the additional revenue, small as it was, helped to decrease slightly the ever-rising deficits of the postwar years.

Subsidiaries Wound Up. As the L&WV struggled to maintain and improve its operations, its two corporate stepchildren were nearing the end of their existence. The end of the war found the Laurel Line Transportation Company operating a fleet of seven buses, numbered 30 through 36. The oldest was a 1937 Model 70 Ford; the newest a 1945 Model 59B Ford; all the vehicles seated 27. Three trucks, numbered 3, 4 and 5, with a capacity of 1.5 tons each, provided pickup and delivery service. They were built between 1936 and 1941.

The fleet was updated as newer equipment became available. Only two were acquired new; however, buses 37 and 38 were purchased from Ford late in 1946 for $6,914 each. Numbers 39 through 42, all Ford models of 1944 or 1945, were bought secondhand, the last two in 1948 from the Frank Martz Bus Company for $2,500 each. Older buses were sold or retired as new ones arrived, since the company needed only six vehicles to maintain its schedules. A new Ford truck, number 6, arrived in 1947.

Bus revenues peaked in 1947, when $113,985 was earned; by 1950, only $90,949 was realized. At the same time, operating expenses were increasing, changing the 1946 profit of $6,711 to a deficit of $9,795 in 1947, and a loss of $17,948 in 1950. Not even contracts with school districts, such as one with Moosic in 1949, could erase the red ink. The L&WV charged an average of 20 percent of its officers' salaries to the transportation company, which was a reasonable ratio, based on the amount of time spent on its affairs.

Less than a year of rail express service remained as car 202, one of the original fleet ordered in 1903, races south along the Susquehanna River in Plains Township during the middle of the last week in July, 1947. The Prospect Colliery of the Lehigh Valley Coal Co. looms in the background. Today only traces of the once booming anthracite industry remain in Lackawanna and Luzerne counties.
Edward S. Miller

A General Motors 33-passenger bus still carries the Laurel Line name as it sits curbside on Linden Street at Adams Avenue, Scranton, awaiting a run to Pittston in 1964. The photographer is standing on the edge of the Lackawanna County Court House square. The L&WV sold its local bus operations in 1951 to the Price Bus Co., which continued to use the old name.
Motor Bus Society, Inc. Collection

With the parent railroad company hard-pressed for cash, and with no prospects for improvement by the transportation subsidiary, the L&WV began to withdraw from highway services. The first step was taken in April 1948 when the trucking franchise and vehicles were sold to Royal Motor Express of Scranton. Rail express service was discontinued on April 14, 1948, with car 202 making the last run. Three years later, the bus franchise for the Scranton–Wilkes-Barre route, on which the Laurel Line offered only minimal service, was sold to Greyhound for $25,000. The much busier local run between Pittston, Moosic and Scranton, was sold to the Price Bus Company for $20,000. The new owner continued to use the Laurel Line name until 1972, when the route was taken over by the publicly owned County of Lackawanna Transit System (COLTS). At this point, the proud old Laurel Line name disappeared from the transportation scene.

While the Laurel Line was beginning to curtail its highway operations, changes were underway on its other subsidiary, the Lackawanna and Wyoming Valley Power Company. In September 1946, the power company sold its coal washery at Rocky Glen to the Gravity Coal Company for $10,000. This firm was in the business of cleaning silt and ash, available from culm banks in the area, and after processing, selling it to manufacturers of cinderblock, then much in demand. The L&WV benefited, since it carried silt from the Butler culm dump on the Erie Railroad to Gravity Coal and was the originating carrier on shipments of cinders to off-line manufacturers. Unfortunately, the quality of Gravity's product deteriorated in 1948 and eventually the enterprise failed.

Such revenue as the deal with Gravity Coal provided was far from sufficient to make the power company profitable. As early as 1947, Watters began to weigh the advantages of closing the power plant and converting to purchased power. The machinery in the plants and substations was old and expensive to maintain; power was produced at the obsolete 25-cycle service, whereas commercial plants generated energy at 60 cycles; the number of customers served by the L&WV Power Company had declined to a handful, with no prospects of gaining more. In 1948, electric service to the Ewen Colliery of the Pennsylvania Coal Company was discontinued and the Ewen substation was closed.

Even though hours and the laboring force at the power plant and substations were reduced in 1949, at a saving of more than $7,300, losses mounted. A careful study was made in 1951 of the benefits and costs of conversion to commercial power. It was estimated the line would save about $55,000 a year by purchasing power from the Scranton Electric Company. If the power plant building could be rented to an industrial firm, then additional revenue might be gained from freight shipments.

The decision was made in April 1951 and approved by the appropriate authorities (the company then being in bankruptcy). The line had to purchase three reconditioned converters for $25,000, install them, alter the signal system and make other modifications. The remaining commercial customers were informed that "prohibitive" costs would compel the power company to abandon all operations. Among them were the John Conlon Coal Company, Number 9 Coal Company, Rocky Glen Park and about nine small businesses in Scranton.

Although the changeover was supposed to take place about November 1, 1951, it was not until July 2 of the following year that the L&WV power plant shut down, and electricity generated by the Scranton Electric Company began to flow through the transmission lines. The final cost of the conversion was $60,238, and 18 men lost their jobs.

In light of the shaky financial position of the L&WV, the expensive switch to commercial power must be questioned. Ironically, less than a year after the decision was reached to make the change, the line applied to abandon its passenger service, which ceased at the end of 1952. Electric freight continued less than eight months beyond that date. Thus, the company had no chance to recoup its investment in its new converters and transformers.

Decreasing Service, Increasing Expenses. Despite the decline in passenger loads which began in 1947, levels of service remained fairly constant through the remainder of the decade. About 37 trains were operated on weekdays, offering 30-minute headways from 5:00 A.M. to 8:00 P.M. and hourly thereafter until midnight. Sunday headways were 60 minutes from 6:00 to 10:00 A.M.; then 30 minutes until late evening. A nightly owl car wended its lonely way along the rails every 90 minutes. Summer service was increased slightly by maintaining 30-minute headways until midnight, although this practice was discontinued in 1949.

Strangely, one limited train, to accommodate Veterans Administration workers in Wilkes-Barre, was scheduled each weekday afternoon beginning in 1946, but it did not appear in the public timetables until 1950. In that year, Sunday runs were scheduled on an hourly basis. The biggest cuts were made in September 1951, when 10 trains were removed on weekdays, which now had hourly service except between 2:00 P.M. to 6:00 P.M. when 30-minute headways prevailed. Sunday runs were slashed from 33 trains to only 12 trains, with hourly headways between 1:00 P.M. to 6:00 P.M., and two hours at other times. The owl car was discontinued.

These reductions were necessitated by a decrease of about 78 percent in passenger loads between 1947 and 1951. Revenue declined about 46 percent in the same period, since fares were increased about 15 percent in 1947 and again in 1948, and about 10 percent in 1949 and once more late in 1951. Occasional charter trips by railfan groups, particularly in the line's last years, did little to change the downward trend.

Periodically, studies were undertaken to analyze the road's costs and prospects. The conclusions were usually gloomy. In 1948, an analysis revealed that the cost per car mile for freight was $1.07 and for passenger trains 67¢, for a combined average of almost 76¢. The writer noted:

> From the above, using the distance of 18.96 miles between Scranton and Wilkes-Barre, the cost of a round trip for a passenger car (37.92 miles) amounted to $25.36 in 1948. Since the average fare collected per passenger in 1948 was .229 it would take an average of 111 passengers on round trip per car to meet operating expenses alone, with no consideration taken for any interest, etc.

Few L&WV trips by this time carried sufficient loads to meet this criteria, so operating losses added significantly to the company's deficit. In 1950, an extensive comparison was undertaken between the L&WV and the Baltimore and Annapolis Railway, another electric road of slightly longer

Although passenger loads were falling after the war, Rocky Glen Park still generated considerable summer traffic. This two-car train was required to haul this group of fun-seekers on Independence Day, 1947. Motorman Fred O'Brien relaxes while conductor Jesse D. Caryl, another veteran of the early days, assists passengers climbing down onto the cinder platform at the park. *Edward S. Miller*

Special events at Rocky Glen Park, such as "Italian Day," "Lithuanian Day" and "Nickel Day," as well as some church or fraternal club picnics, sometimes required unscheduled trains to be operated between Scranton and Pittston. One of these extras, comprised of coaches 34 and 39, crosses over from the southbound to the northbound track in Pittston in order to return to Scranton on Independence Day, 1946. The trainman stands ready to restore the switch for through traffic after this movement clears the points.
Edward S. Miller

The Wilkes-Barre station site in 1950, with one car loading on the loop and a spare on the station spur, bears no resemblance to the area today. The Wilkes-Barre *Sunday Independent* is published in a printing plant recently erected here. *William D. Middleton*

mileage, which ran between its namesake cities. The survey revealed the B&A cost less to operate, due to lower wages, no stations, and a conscious decision to allow the property to deteriorate in the face of a forthcoming passenger abandonment petition. B&A buses competed directly with its cars, and new housing projects gave some promise of profits from highway service. A frustrated L&WV official wrote, "The Laurel Line is practically dead as far as new housing . . ." is concerned.

The Laurel Line Goes Bankrupt. Until the postwar era, the L&WV had led a charmed life so far as maintaining its financial independence was concerned. Although it was in a perilous condition several times, and was, in fact, technically bankrupt, its bondholders had an unusual sense of forebearance and the line always managed to cling to its corporate freedom. This situation was now about to change, as the old magic no longer seemed to have any effect.

At first, the situation seemed hopeful. The new management was competent and optimistic. Watters was a specialist in reorganizing failing companies and he regarded the little electric road as a challenge to his skills and abilities. Nor did he wish to end passenger service, since it still provided about 75 percent of the company's revenues.

One of Watters' first steps was to soothe the long-neglected stockholders, who rarely heard from the company and who had received no dividends since 1930. Periodically, a few of these unfortunate investors wrote letters to the firm, complaining of the lack of attention. Said one, "I have felt for years that the stockholders deserved to be kept better informed of the company's affairs," while another wrote, "I have held the stock many years and have never received even one cent."

So dismal were the line's prospects that when one man tried to sell his shares in September 1947 he could secure no better than 3/8 for the common and 1 for the preferred. It is no wonder that, as a group, the stockholders were apathetic, and every year the line had to beg and plead to get enough proxies to secure the majority necessary to transact business. Many of the shares were scattered in small lots, while others were held by the estates of deceased investors. The company's records were sometimes deficient too; in 1947 it was discovered that one estate listed as holding 802 shares had sold them at auction in 1930!

Although he recognized the stockholders had no role in determining the company's fate, Watters felt they were entitled to information if not dividends. He therefore decided to issue an annual report for the year 1947, which would be the first time the company had undertaken to release such a document.

It was not a particularly encouraging time to do so. Not only had passenger revenues begun to slip but the expenses and the deficit rose alarmingly. Anticipating a flood of queries from the now-aroused stockholders, Watters told Treasurer J.J. Loftus to prepare an explanation in the annual report, centering on the rising cost of wages and materials, the need to catch up on deferred maintenance and the lag in fare increases. Noting that the mighty Pennsylvania Railroad was also experiencing difficult times, the L&WV president sighed, "It would seem to me that 1947 was a particularly hard year in which to operate a railroad."

The final report, which no doubt produced more surprise from the stockholders than anything else, amounted to 12 pages, which briefly described the road's background, finances and prospects. The reading was not pleasant, as the net deficit for the year amounted to $152,170. The experiment of an annual report was not repeated, since by the following year the line was plunging rapidly toward receivership.

As Watters was well aware, the real problem, which seemed to defy solution, was the soon-to-mature bonds and debentures, which, together, represented an obligation of almost $3 million. The debt consisted of $1,950,000 of first mortgage, five percent bonds due August 1, 1951, and $999,500

Pulling an unusual mix of cars, northbound locomotive 401 ascends the grade above Pittston station along the Erie Railroad on Thursday, June 24, 1948. Included in the consist are L&WV ash car 716, Armour Refrigerator Line 1691, a Northwestern Refrigerator Line car, a Merchandise Dispatch reefer, and Laurel Line boxcar 505.
Edward S. Miller

of 20-year, six percent debenture bonds, due February 1, 1948. Unpaid interest on the securities totaled $623,340, and accumulated, unpaid dividends on the preferred stock added $1,225,000. There was no way the L&WV could meet these obligations. Watters was forced to pin his hopes on a financial reorganization which would wipe out much of this debt and give some promise that the carrier could earn its keep thereafter. Regrettably, circumstances developed to make this hope futile.

The key to Watters' plan was a proposed amendment to the Interstate Commerce Act which would provide for voluntary readjustments of railroad debt structures when three-quarters of the holders of the securities were willing to accept the necessary modifications to avoid bankruptcy proceedings. By the time this legislation, known as the McHaffie Act, passed in 1948, the L&WV bondholders had organized several "protective committees" to safeguard their interests.

Late in 1947, when the bill was still pending, Watters wrote a form letter to the securities holders, pointing out that the railroad was ". . . carrying a capitalization which is much too large for its earning capacity" and pleading for patience. As soon as the law passed, Watters called a conference of all the interested parties and worked out a plan of reorganization. Under this proposal, the maturity date of the first mortgage gold bonds would be extended from August 1, 1951, to August 1, 1996, and the interest on them reduced from five percent to two percent, with the remaining three percent contingent on earnings. The mortgage would be modified to permit the issuance of new equipment bonds as a first obligation. In similar fashion, the maturity date of the six percent gold debentures would be extended from February 1, 1948, to February 1, 1993, and the fixed interest rate changed to one percent contingent on earnings.

On August 2, 1948, the L&WV filed an application with the ICC to modify its securities in the manner described. Briefs were filed, a hearing held, and an examiner issued a report on May 11, 1949, recommending denial of the application, since he felt that even with the modifications, the L&WV would be unable to earn an amount sufficient to pay the reduced interest. Two days later, a suit was filed by a bondholders' protective committee in the U.S. District Court asking for a financial reorganization under the traditional section 77 of the Bankruptcy Act. The petition was approved on June 24, 1949.

For the first time in its almost half century of existence, the L&WV was legally bankrupt and under the protection of the courts. After a hearing, and considering letters from interested parties, the court appointed E. McLain Watters as trustee, a decision which was ratified by the ICC on November 10. While the bankruptcy relieved the pressure on the line from the security holders, it did little to offset the problems generated by the rapidly declining passenger revenues. One of the problems which Watters now had to face was how to satisfy the demands of labor.

Intimidating to any motorist is the sight of train 5 crossing North Main Street in Wilkes-Barre. The automobile manufacturers required three model years after the war to fill all of the orders resulting from the curtailment of production during the four war-time years. This shortage of private transportation required the Laurel Line to assign both cars 38 and 36 to this train. Alas, within six more years, there would be no demand for the railway's services and today most of the right-of-way is used to park automobiles. *C.L. Siebert, Jr.*

Leading six loaded hopper cars, locomotive 401 (ABOVE) heads south over the Avoca viaduct which spanned the Lehigh Valley Railroad, the Delaware and Hudson Railroad, and U.S. route 11. By this time, June 14, 1948, anthracite trade was falling sharply and would soon propel the L&WV into bankruptcy. *Edward S. Miller* Before this crew (AT LEFT) can retire after a long day, it will have to work the train forward toward the Scranton Terminal, then back up the remnant of the Dunmore Branch, and then enter the freight yard. *Charles R. Yungkurth*

The Railroads Had it Rough, Too.

The post-war traffic boom was still in full swing on Easter Sunday, 1946, forcing the Lehigh Valley Railroad to make up a twelve car extra passenger train which ran as the first section of train No. 10, "The Black Diamond," from Pittston and Wilkes-Barre to New York. This action prevented overcrowding of the regular Buffalo to New York train, which is entering the Wilkes-Barre station immediately behind this train. Fifteen years later the station closed.
Edward S. Miller

The Central Railroad of New Jersey offered train service to Philadelphia, New York, Allentown and Reading from its terminal on the west bank of the Lackawanna River in Scranton. Not only is its Scranton station extant, but the Wilkes-Barre station, opposite the former home of Stegmaier's Gold Medal Beer, is now a motel and restaurant complex. Illustrated in 1940 is the carrier's first morning down train to Philadelphia crossing the Lehigh River in Bethlehem, 2 hours and 57 minutes from Scranton and a mere 1 hour 47 minutes from Philadelphia. Unable to compete with faster service on the Lehigh Valley, the CNJ dropped daylight varnish in 1950. *John J. Bowman, Jr.*

The New York, Ontario & Western Railway was the third trunk line, in addition to the Delaware & Hudson and the Erie, with which the Laurel Line would have competed had its proposed Scranton to Carbondale line been built. Shown in 1932 is an NYO&W coal train at Cadosia, New York, where the Scranton Division entered the mainline. Dependent on coal trains and summer passengers to the Catskill resorts and suffering from stronger railroad competition everywhere it ran, even the economies of early dieselization were insufficient to keep the NYO&W out of bankruptcy. The "Old Woman" ceased in 1957 and today one of the few surviving landmarks are the bridge abutments flanking route 6 at the south edge of Carbondale.
Thomas J. Flanagan Collection

The era of the passenger train for the "public necessity and convenience" in northeast Pennsylvania lasted 139 years and ended with the departure of Erie-Lackawanna Railway's "Lake Cities" for Hoboken and New York in the frigid early morning of January 5, 1970. Vanishing into the night along with the markers was a long tradition of immaculate accommodations, impeccable service, and food, not warmed up, but cooked to perfection in the dining car. In a more prosperous era at the end of World War II, the "Lackawanna Limited," the flagship of the carrier between Hoboken and Buffalo, is shown entering Scranton's station precisely on-time amid a cloud of brake smoke after its long descent into town. Included in its consist that day were coaches for Buffalo, Chicago, and Oswego, a parlor car, a diner, one sleeping car for Chicago, and a combination baggage and Railway Post Office car. When Scranton was at its zenith in 1930, 28 trains a day stopped at this station, situated next to the Laurel Line facility. *Edward S. Miller*

On a sunny January morning in 1946, Delaware and Hudson Railroad's engine 500 steams up to the station at the west end of Lackawanna Avenue in Scranton, today the site of the Greyhound terminal. Train 510 had originated in Carbondale and was among the last vestiges of D&H's once extensive passenger service in the Wyoming Valley. Six years later, with mine employment on the skids and residents moving out of the valley in droves, the D&H would throw in the towel. *Edward S. Miller*

Framed by the western mountains, the Susquehanna River, and the Pittston business district, combine 115 accelerates northward from the station in the final month of passenger service. Passenger trains on the Lehigh Valley Railroad, which stopped in Wilkes-Barre, ran parallel to the near bank of the river through Pittston.
Edward S. Miller

The Laurel Line's Only Strike. An indication of the pressures and tensions developing on the faltering line was the worsening relationship between management and labor. The postwar period was marked by annual demands by the labor unions for wage increases. Although the L&WV refused to be bound by the national agreements reached with the Class I carriers (those railroads which grossed more than $1 million a year), it nonetheless granted raises of about 18 cents an hour in 1946 and 15 cents an hour in 1947. In the following year, however, with the line drifting rapidly toward the financial rocks, Watters claimed the company could go no further. Trainmen continued to earn $1.27 an hour while Watters tried, unsuccessfully, to avoid bankruptcy. Eventually, in 1951, a raise of 6 cents an hour was granted, but the issue of back pay remained.

Adding to the acrimonious air between labor and management was the company's contention that some employees were cheating by permitting favored passengers to ride free. In a memo to Superintendent John Hennigan, Treasurer Loftus, stressing that only a few trainmen were involved, stated that "a great number of commutation tickets have been punched over and over again in excess of the number of rides allowable under the ticket." Other tickets were accepted as much as a month after their expiration date had passed. With some exaggeration, Loftus declared, ". . . if we had revenue from all such excess we wouldn't be worried about meeting our payrolls and other operating expenses." The union naturally resented these charges, although evidently they had an element of truth.

By 1951, the patience of the Brotherhood of Railroad Trainmen (BRT) was wearing thin. W.L. Reed, assistant to the president of the BRT, complained that some nonoperating employees were receiving more per hour than his men and that the line had evaded the payment of back pay. Watters, now trustee of the bankrupt road, denied the charges and stated the financial condition of the carrier was ". . . at the lowest ebb in its history. For the first time, we have not had sufficient cash on hand to meet the semi-monthly payroll and the drafts of foreign roads upon us for prepaid freight shipments." He blamed this distressing state of affairs on strikes at General Electric and Harris-Hub and delays in converting to commercial power. His parting shot was a warning that the line was faced with the possibility of abandoning passenger service.

Undeterred by this implied threat that most of the men might lose their jobs, angry over their inability to collect the

$21,000 in back wages due them and demanding a raise of $2.00 a day, the 30 motormen and conductors walked out at 2:00 A.M. on December 19, 1951. The line immediately shut down, and about 60 nonoperating employees were furloughed.

Newspapers in the region deplored the strike, the *Wilkes-Barre Independent* asking, "Would they strike if the sun were shining?" Since only about 2,400 rode the trains daily, the region was hardly greatly inconvenienced. A week after the strike began, on December 26, Watters asked permission of federal judge Albert L. Watson to abandon passenger service. A hearing was scheduled for late in January. The competing Price Bus Company, operating the Laurel Line Transportation Company's former route between Pittston and Scranton, increased its service between those points, thus further damaging the prospects of the electric line.

As with most strikes, the negative impact outweighed any positive gains. Fortunately, both sides soon recognized this. Prodded by Mayor James T. Hanlon of Scranton, who offered his good offices as a mediator, Watters, Reed and other officials met on January 7 and in one day resolved their differences. The agreement of January 8, 1952, provided for a 10¢ per hour increase, effective March 1, but not payable until July. Two months' back pay would be paid by May, and the balance would be paid in a time and amount to be determined by later negotiations. The next day the strike ended and service was soon restored.

The strike, which lasted three weeks, came at an unfortunate time for the carrier. Watters' earlier optimism that the line could continue as a passenger carrier had by now been completely dashed. The labor dispute marred the line's long record of generally benevolent relations with its employees. The underlying cause was the eroding revenue base and the growing sense of frustration on the part of both management and labor. While it is unlikely that the strike directly resulted in the plan to abandon passenger trains, some thought it was a case of cause and effect. The petition to abandon was not withdrawn, and the line now turned its full attention to the removal of passenger service, in what would be a final effort to survive as an independent railroad.

Looking unusually refreshing is this silhouette of a car passing the Pennsylvania Coal Company's Number 6 Pond at Ewen. Compare this picture with that on page 172 showing the adjacent culm banks. *Edward S. Miller*

Photographer Edward S. Miller stood on top of a cliff at the site of the old Meadowbrook Crushed Stone Company to record coach 36 crossing Stafford Meadow Brook enroute to Scranton on March 18, 1952. Due to this area's isolation, it is unchanged save for the construction of Interstate highway 81.

This montage includes two elements of Wyoming Valley history: the Laurel Line interurban car and the Ewen Colliery property of the Pennsylvania Coal Co. Although the volume of coal mined diminished rapidly after the war, the end of deep mining in the valley was abrupt. Miners working in the Knox Mine, not far from the Ewen Colliery, accidentally broke out into the Susquehanna River on January 22, 1959, resulting in the loss of 12 lives. Because all of the mines were interconnected underground, it rapidly became uneconomical to pump water from any of them. Only seven years after the Laurel Line closed its ticket windows, deep mining had also vanished in the Valley. A few small breakers remain to process strip coal; the last large one, in Ashley, has not worked since 1982.

Edward S. Miller

In sharp contrast to the grim mining environment on its southern end, the northern section of the right-of-way often appeared in a bucolic setting. Sporting the vermilion red with blue stripe color scheme, introduced in the Spring of 1949, number 39, running northbound, has just discharged a pair of passengers at the Moosic station, situated at Spring Brook Road, designated as Pennsylvania highway 502. The original Moosic station was located south of here at Powder Mill Road.

Edward S. Miller

This ramschackle breaker was one of the few served by the Laurel Line. Southbound combine 116, with its baggage compartment leading according to customary practice, passes the colliery located below Rocky Glen Park in the Borough of Moosic, October 4, 1952. The facility had several owners but was generally known only by their initials, such as B&J or P&J. From 1850 to 1885 the Pennsylvania Coal Company Gravity Railroad's number 4 incline plane ended behind the hopper cars in the colliery siding.
Edward S. Miller

The attentive motorman glances at the photographer as southbound coach 32 moves over the cinder-strewn right-of-way along the Erie Railroad below Moosic station. It was 2:45 p.m. on a sunny Saturday, October 4, 1952. *Edward S. Miller*

The empty structures of Rocky Glen Park in Moosic await the onset of winter as combine 116 passes the deserted platform. The autumn leaves reflecting in the lake are prophetic, for next year private automobiles would bring all patrons to the park. *Edward S. Miller*

A gentle snow is beginning to coat the right-of-way as 114 bursts from the south portal of the Crown Avenue tunnel late in the morning of the last Saturday of passenger service. *Edward S. Miller*

Passenger Service is Abandoned. Receivership provided a little breathing space for the struggling L&WV, but it could not stem the decline in revenues, especially from the passenger service. Railway operating revenues of $676,843 were reported in 1947, but taxes, interest and expenses resulted in a deficit of $152,170. There was a slight improvement in 1948, caused by an upsurge in freight traffic, but still a deficit of $42,876 was posted. The year 1949 marked the beginning of a sharp decline, as passengers carried fell to 1,573,386 and the deficit jumped to $141,665. Even worse was 1950; the passenger count was 1,050,581 and the deficit a frightening $354,782.

Things were so bad that the company was forced to ask the Lehigh Valley and Lackawanna Railroads to delay presenting their drafts for interline freight charges due them since there was no cash on hand to pay them. Only court approval of the issuing of $100,000 in trustee certificates prevented a complete shutdown.

In the face of this financial storm, Watters did what he could to stimulate interest in the passenger trains. Free parking was offered at some stations, but few took advantage of it. New cars were out of the question, and Watters never seriously considered converting the heavy, steel cars to one-man operation, partly because the union would strenuously object to such a move.

One thing he could do, relatively cheaply, was to spruce up the present equipment with a new paint scheme. The first effort in this direction was less than successful. Coach 32 emerged from the shop in February 1949 with a vermilion body and letterboard, broken by what was described as ". . . a deep gaudy blue" along the belt rail and window posts. Even worse was the interior, where the dark mahogany had been painted a battleship gray. The local press and commuters derided the attempt. One asked, "Is this the one that you get in a box of corn flakes?" while another asserted that coach 32 was ". . . gaudier than any cars in the train of the Ringling Brothers Barnum and Bailey circus." Others said it looked like the inside of a hospital.

Subsequent repaintings were better, beginning with car 37, which retained the vermilion exterior, now enhanced by two attractive blue stripes above and below the windows, while the interior retained its natural mahogany finish. Gradually, the other cars were repainted, except for 35 and 36 which kept their tuscan red until the end.

But fresh paint could not lure the public off the highways and back on the electric cars. In 1951, the year shortened by the strike in December, a mere 882,618 fares were collected, and the deficit was $260,178. To Watters it was now obvious that the only way to save the line from collapse was to abandon the passenger service. On December 26, 1951, the line asked permission of federal judge Albert L. Watson to apply to the regulatory agencies for the right to discontinue the passenger trains.

Hearings and appeals occupied most of 1952. As is usual in such cases, public officials quickly protested the move and argued the line was necessary for the economic health of the region, even though fewer and fewer rode it.

The initial hearing was held before attorney Julius Levy, special master in railroad proceedings for the Middle District Court, on January 28, 1952. W.L. Cairns and other officials presented evidence of the decline in passengers carried and the line's precarious financial condition. The prime reason cited was the private automobile, against which it was impossible to compete. Although he did not refer directly to the collapse of the anthracite industry, Cairns stated that thousands of miners used to ride, but now only about 150 did so. A representative of the Railway Supply Company of Buffalo, New York, testified the road had a scrap value of about $212,951, a figure which seemed to be considerably underestimated.

On February 26, Levy recommended to Judge Watson that the carrier be given permission to apply to the ICC and PUC for abandonment. Levy wrote, "The estate is simply wasting away and a review of the financial state seems to indicate that it is all due to the failure of passenger service." He exonerated the Watters management from any fault and stated that the reorganization must hinge on freight service. Although Greyhound, the Laurel Line buses, now owned by Bert Price, and Wilkes-Barre Transit buses provided alternate service, officials from Scranton, Moosic, Rocky Glen Park, Avoca, Pittston, Dupont and Wilkes-Barre vowed to fight the move. Judge Watson approved the recommendation on March 12 and by the end of the month, Watters applied to the Public Utilities Commission in Harrisburg for approval.

The Public Utilities Commission conducted a hearing on May 29 in the Lackawanna County Court House in Scranton, which was attended by about 50 people, most of them protesting the planned end of passenger service. They included representatives from the on-line towns, the operators of Rocky Glen Park and the railroad Brotherhoods. Watters claimed the line lost about $200,100 a year from carrying passengers but earned a profit of about $35,000 annually from its freight operations. He argued the existence of the company would be in jeopardy if the carrier were forced to continue the passenger trains. He presented the results of a station-by-station survey of southbound passenger counts taken on May 6 and May 7. Only about 850 passengers boarded the trains in Scranton and the other points. Not surprisingly, about half purchased their tickets in Scranton and about 100 in Pittston. Most of the remaining stations handled fewer than 25 each, while Virginia and Midvale recorded none.

While the PUC held further hearings and deliberated, Watters received an offer to purchase the line's passenger equipment in order to continue the service. John R. Reap, Jr., an attorney representing the Moosic Chamber of Commerce and other undisclosed backers, offered $150,000 for one track and the rolling stock, plus $5,000 for the franchise and $5,000 annual rental. The proposed Scranton–Wilkes-Barre Rapid Transit Corporation planned to secure a loan from the Reconstruction Finance Corporation to finance the purchase.

Watters, after making ". . . an exhaustive study," rejected the offer. Not only would the bondholders disapprove, but the offer of $150,000 for the cars, wires, parts and other facilities was less than the scrap value. Rental of the Scranton and Wilkes-Barre stations alone would exceed the $5,000 suggested as rental for all facilities. Watters was also disturbed by the failure to disclose the principals behind the scheme and the uncertainty of securing the federal loan. Finally, the L&WV trustee stated that the operating problems were ". . . practically insurmountable."

Trying to run passenger trains on the northbound track, as proposed by Reap, and freights on the southbound track, posed many difficulties, nor was it possible to accurately divide power, signaling and maintenance costs among two operators on the same railroad. In short, the proposal was impractical. No more was heard from the abortive Scranton Wilkes-Barre Rapid Transit Corporation.

As passenger counts fell, especially on weekends, Watters made one final timetable change. Effective October 12, 1952, only eight round trips were operated on Sundays, with one car running every other hour between 9:00 A.M. and 11:00 P.M. About the same time, a new crossover was installed between North Franklin and North Main streets on the canal right-of-way, permitting single-track operation between that point and the Wilkes-Barre terminal.

On November 11, the PUC authorized the abandonment of passenger service on 20 days' notice. Watters decided to terminate operations with the end of the year, December 31, partly not to inconvenience those traveling over the Christmas holidays. In its decision, the PUC noted a 53 percent patronage decrease and a 40 percent revenue decrease between 1946 and 1950, and stated there was "an abundance" of substitute services between the two cities.

As the final days passed, the line received unusual attention from the press and public. Newspapers lauded the long service of the electric line but generally accepted the demise

The declining state of the carrier's affairs is grimly evident in this view of the deserted station platform in Scranton only ten minutes before car 116 departed at eight o'clock on December 23, 1952. Eight days later, this car was included in the final passenger train. *Edward S. Miller*

of the big passenger trains as another evidence of "progress." One paper speculated that, if freight service also ceased, parts of the right-of-way would make ". . . a super duper highway." Some pointed out that on Memorial Day, 1924, when John Mitchell's statue was unveiled in Scranton, the line had hauled a record 72,344 people, in sharp contrast to the 1952 Memorial Day, when only about 3,000 passengers were handled. A number of charter trips were operated, including a "farewell" run by the Electric Railroaders Association on August 3 and a final charter by two local railfans, Edward S. Miller and William F. Ellis, on December 28, with about 10 people enjoying the ride on car 39.

All too quickly, the last day, December 31, 1952, arrived. More tickets were sold that day than the previous three combined, as many people availed themselves of the last opportunity to ride the electric cars. The final run, consisting of combine 116 and coach 38, left Scranton shortly after 10:00 P.M. The crew consisted of motorman William Kapp, and conductors John A. Ratchford and P.J. Grier, all well up on the seniority list. Aboard were General Manager W.L. Cairns, Superintendent J.J. Hennigan, other officials, longtime active and retired employees and W.A. Druffner of Avoca, who had ridden the first run to Pittston on May 20, 1903.

About 170 passengers crowded onto the train, which arrived in Wilkes-Barre about 10:45 P.M. After turning on the loop and loading the last mail sack, the train departed at 10:54 P.M., nine minutes late. At 11:41 P.M. the train pulled into the Scranton terminal, the passengers slowly departed, and motorman Kapp ran the train into the yards, thus closing almost a half century of passenger service to the citizens of the valleys.

Bids were solicited for the nine coaches and three combines no longer needed by the L&WV. On May 11, 1953, the bid of $12,000 by John Bauman of Carbondale and Sam Landau of Paterson, N.J., as partners, was accepted. Oddly, the cars remained in storage outside the Scranton shops until April 1954, when they were finally removed to the NYO&W's Mayfield yard, adjacent to the Bauman scrap yard, near Carbondale, where they remained until 1956, when scrapping finally began. By April they were gone.

The end of passenger service marked a drastic change for the L&WV. Continued electric operation was doubtful, and much surplus equipment and plant faced retirement. It remained to be seen whether a stripped-down Laurel Line could survive as a freight-only carrier, or whether complete abandonment of the once busy electric road was likely.

As its headlight pierces the darkness, combine 116 prepares to leave Pittston station, southbound, on Tuesday, December 30, 1952. The hour was 10:22 p.m. and exactly twenty-four hours remained until the last southbound passenger train would leave. The crossing watchman left his lantern on the street while he labored on the opposite side of the car to load mail.

Edward S. Miller

The Last Passenger Train

The life of most interurban electric railways was so brief that many employees saw the line come and go. The Laurel Line was no exception. Posed for a photographer from the *Wilkes-Barre Record* immediately before the last train out on December 31, 1952, are many long time employees and friends. From left to right on the platform are Jesse D. Caryl, a retired conductor who started in June, 1906; Superintendent of Transportation John J. Hennigan, who started as an office boy in 1903; Michael J. Lavelle, a railway hobbyist and trolley operator from Washington, D.C.; and Joseph P. Fitzpatrick, a passenger who had previously purchased the first ticket sold in Pittston in 1903. Also included are General Manager William L. Cairns, a career employee with the L&WV since 1903; Wilkes-Barre General Freight and Passenger Agent George W. Bainbridge, who worked for the line from 1905 until 1960; and L&WV Power Co. Chief Engineer George Dewey Shrader, who began work in 1916 and continued under Erie-Lackawanna management. The train crew included Conductor John A. Ratchford, whose seniority dated to 1911; and Motorman William E. Kapp, who had worked since 1908. The tenth person is an unidentified passenger. The youngest employee in the picture is the last in line, Patrick J. Grier, with only five years seniority. Inside the second window from the left is Robert Francis Boland, a former motorman who worked from 1905 until 1943. He had previously been a steam locomotive engineer with the contractor who built the L&WV. Boland's son also worked for the Laurel Line and retired as conductor before the Conrail takeover. Inside the third window is Charles Dorsey, a retired conductor who started in 1907.

Edward S. Miller Collection

On June 1, 1969, Billy Smerdon, John Weselenko and Edgar Boland, the frustrated crew of a southbound freight, reflect upon a heat kink below Number 14 station. A planned rehabilitation of the line by its new owners was never accomplished. By this time, only about one crosstie in ten was acceptable. Extreme heat caused the rails to expand and, with nothing anchoring them, they simply buckled. *Edward S. Miller*

7

The Final Years: Diesels, Freight and Merger

SHORTLY AFTER the L&WV passenger cars turned their last revenue wheel, a dramatic change in administrations took place in Washington. In the November elections, the voters had chosen Republican Dwight D. Eisenhower over the Democratic candidate, Adlai Stevenson. In January 1953, the former World War II hero took possession of the White House, while his party controlled Congress by a bare majority.

The popular president, who was reelected in 1956, maintained a low-key approach to his office. Among his first achievements was the negotiation of a shaky truce in July 1953, which ended the Korean War and which was still in effect as of this writing 30 years later. A summit conference at Geneva in 1955 brought a welcome, if temporary, respite from the pressures of the Cold War, although trouble spots in such places as Formosa, Indochina and the Middle East kept Americans alert and uneasy.

At home, the Eisenhower years saw a renewed drive to extend civil rights to blacks, the launching of satellites into outer space, the downfall of Senator Joseph McCarthy, debates over private versus public power, and the passage by Congress of the Federal-Aid Highway Act of 1956, which provided for the construction of more than 41,000 miles of interstate roads and highways, at a cost then estimated at $33.5 billion. Although this latter act would eventually have a drastic negative impact on the nation's railroads, it had little effect on the Laurel Line, which was now struggling to survive in a form far different from that conceived by its optimistic promoters at the turn of the century.

The Laurel Line Retrenches. With passenger service only a memory, the Watters management began to partially dismantle the once-impressive electric line. Freight would require only one track, the signal system was now redundant and even the continued use of electric power was highly questionable.

One of the first steps was to dismiss most of the remaining crossing watchmen, since the occasional freight could be flagged over the few streets in Wilkes-Barre and Pittston. An odd, temporary arrangement was made in the latter town, where local officials and the PUC demanded watchmen remain at Market and Pine Streets because the third rail was judged to be even more dangerous than before since so few trains ran that children or careless adults might forget its hazard and risk electrocution.

The line's answer to this problem was to erect a trolley wire in July over a short stretch of track through the town, thereby enabling it to remove the watchmen and deenergize the third rail. This temporary expedient lasted only a month, since diesels were to replace the electric locomotives in August.

Also deemed superfluous was the signal system, which had been installed after the 1920 Pittston disaster, as only one freight a day was anticipated. Application was made to the Interstate Commerce Commission (ICC) for permission to remove the signals. Approval was duly received in two stages. The first was taken on April 21, 1953, when signals between Virginia, south of the Scranton tunnel, and Wilkes-Barre were discontinued. The remaining signals, between Virginia and Scranton, were cut out on January 20, 1954, subject to the condition that only one train be operated at a time on the line. A total of 35 signals was removed.

Double track was now a luxury the line could no longer afford. Since the majority of sidings were on the northbound track, Watters decided to scrap the southbound rails. A petition was filed in the federal court in Scranton asking permission to remove about 16 miles of track, which was granted on August 26, 1953. A public auction was held in Scranton on October 8, with the winning bid of $225,000 submitted by the Hyman-Michaels Company of Chicago. Included were the running and third rails, insulators, wire and other fittings, extending from Scranton to Wilkes-Barre, although short sections were retained in several places for switching purposes.

Using leased DL&W diesel 51, a caboose, two gondolas and a flatcar equipped with a crane, wrecking crews completed their task before Christmas. Thus finally disappeared the fine, double-track speedway of which Charles Conn and the early promoters had been so proud.

At the same time these decisions were being reached, Watters concluded the electrical plant must also go. Although more than $60,000 had recently been spent to convert the

With the wooden framework of the "million dollar" roller coaster in the background, locomotive 401 hurries a short train north at Rocky Glen Park on June 17, 1953. The two empty refrigerator cars in this consist were typical of a nationwide fleet that was predominantly ice-cooled at this time. This forced the Laurel Line to handle short trains of these cars on a costly priority basis.
Edward S. Miller

Locomotive 403 draws power from the overhead wire, which was erected for a brief period in Pittston to permit deadening the third rail and furloughing the watchman at this crossing. The former southbound track at Pine Street was obviously out of service because only the one-time northbound track had been wired. The date was July 25, 1953. *Edward S. Mille*

The Maffett Street School, in Plains, serves to locate this picture of 403 trundling a single box car northward. It was 12:50 p.m., Saturday, August 15, 1953. Four days later the power would be permanently turned off. *Edward S. Miller*

system to purchased power, it made little sense to maintain the substations and distribution network, plus the three ancient locomotives, for one or two trains a day. Purchasing a replacement diesel was financially impossible and strategically unwise, as Watters was already anticipating a merger with the Lackawanna Railroad as the line's ultimate salvation. Leasing a locomotive from the DL&W would not only help cement relations but also permit the Laurel Line to dispose of its carhouse and shop as well.

An agreement was soon negotiated between the L&WV and the DL&W, whereby the larger carrier agreed to lease a diesel for $50 per 10-hour day, with a charge of $2.40 for each additional hour. Although the L&WV was required to give a $4,000 deposit, the Lackawanna would provide all service and maintenance.

Watters soon raised the necessary cash and on August 19, 1953, DL&W 51, a 380-hp, 44-ton General Electric product of 1948, replaced the electric locomotives. On the same day, single-track operations officially commenced between South Scranton and Dupont. Little 51 proved to be inadequate and was replaced by larger machines, often DL&W 410 or 435. The rental fee was increased to $61 a day, but it remained at that rate for the remainder of the L&WV's independent existence. Standard practice was to pick up the locomotive at about 4:00 A.M. and work until 2:00 or 3:00 P.M. Rental payments averaged about $1,850 a month.

The three obsolete locomotives remained on the property until March 1956 when they were sold to Albert Niebert and Company, a local scrap dealer, for $8,153. Unfortunately, the historical significance of the locomotives was not fully realized. The three machines, including 401, the 1895 Baldwin-Westinghouse pioneer, were hauled to Niebert's scrapyard on the Minooka branch where they were cut up. So ended another phase of the L&WV's history.

In a general house cleaning, the L&WV was stripped of almost its entire roster of equipment. All the cars were obsolete and barred from interchange service, and Watters could see no point in retaining them. Included in the sale to the Fiegleman Waste Materials Company were seven wooden boxcars, four steel hopper cars, eight wood-bodied ash cars, two flatcars, five low-side, wood gondolas, the derrick car, the flanger and the hulks of coaches 19 and 23. It was estimated the cars, which brought $5,968 to the embattled line, contained 512,000 pounds of scrap metal. The body of one boxcar was kept for storage, while another, 502, was used for a time as a work car.

Surplus real estate was put on the market. The Wilkes-Barre station on East Market Street was sold in February 1954 to Alfred Crane of the Crane Taxi Company for $18,000. The building was old and in poor condition, and Crane did little to improve it. By the 1960s, the taxi office moved elsewhere, and the structure became an eyesore, with broken windows, abandoned automobiles, tires and boxes littering the grounds. Under pressure from the city, Crane razed the building in 1972. Some of the land adjoining the freight station was leased to Wilson and Company in 1954 for $770 annually for use as a parking lot.

The former Plains substation and station fared no better. Initially, it was leased for $300 a year to a local businessman to store furniture. Later it was sold and finally it was damaged by fire in 1977 and soon demolished.

The remaining buildings, once so impressive and the source of much pride to the line's early managers, were eventually sold or torn down. The powerhouse stood derelict for some years before it was razed in the early 1960s as part of a road construction project on the old over-the-hill line right-of-way. The Scranton Terminal Station served no purpose after the merger with the Lackawanna Railroad, and it, too, was demolished early in 1967. Also a victim of the wrecking ball was the Pittston passenger station, which was razed to make way for a community swimming pool in 1963, and the Avoca substation, which fell about the same time.

The only surviving structures are the former carhouse in Scranton, which is used as a factory by the Scranton Laminated Plastics Company, and the Scranton and Pittston freight stations, which have been converted to other uses. By the early 1960s, aside from a few leaseholds, all that remained of the L&WV was its weed-covered single track stretching between Scranton and Wilkes-Barre. Even most of this would not long survive under the Erie-Lackawanna and Conrail ownership. About the only positive aspect of all this was that management no longer had to ponder whether the line was an electric interurban railway or "a steam road that just happened to be operated by electricity." At long last, the L&WV's "identity crisis" had been solved.

In a scene which now can be relived only in memory or in pictures such as this, the distinctive L&WV locomotives 401 and 403 rest in front of the car house early on a Sunday afternoon in May, 1953. The smoke rising from a hot journal bearing on the right locomotive implies that immediate attention from the shop force is needed. The normally bewildering management practices of the Laurel Line are evident in the recent rebuilding and repainting of 403, after an accident. It would be idled in only three more months. *Edward S. Miller*

Leased DL&W locomotive 51 works the Wilkes-Barre yard along North Pennsylvania Avenue on Saturday morning, August 29, 1953. An old L&WV box car is on a siding on the right under the now useless overhead wires. All of the buildings were demolished in 1977 by the Wilkes-Barre Redevelopment Authority. The 44-ton diesel was built by General Electric in October, 1948, and was popular for light duty on many railroads because national wage contracts did not require firemen on engines of this weight. On the Laurel Line it did not matter; firemen were never employed. *Edward S. Miller*

The coal loading dock at Number 14 station, Plains, was a major source of freight revenue for the railroad in 1953. The short term overload capacity of the electric locomotives made lifting cars up this grade child's play, but the diminutive 380 horsepower leased diesel was inadequate for such tasks. Today, this location is completely overgrown. *BOTH: Edward S. Miller*

Crewed by the four men specified in the union contract, leased DL&W diesel 51 idles at Plains Station at Perkins Street on Sunday, August 30, 1953. On the footboard are Conductor Edgar J. Boland and Brakeman Thomas Kelly. Brakeman William Deinlein stands on the running board in front of Engineer Joseph Wallace. Substation operator Harold Machler, whose position would be terminated on September 4, hands "orders" to the engineer. *Edward S. Miller*

Freight Traffic Fails to Develop. Although the L&WV realized $282,897 from the sale of most of its assets, it was quickly spent in meeting overdue obligations. More than $100,000 in trustee certificates had to be redeemed, back pay of nearly $49,000 disbursed, various tax claims cleared and foreign road debts to such carriers as the Lehigh Valley, DL&W, Northern Pacific, Southern Pacific and Union Pacific paid. The line could survive only if it were able to earn enough from its freight traffic to pay its operating expenses. Unfortunately, this proved to be impossible.

A policy of close cooperation with the DL&W was instituted. When the large carrier began piggyback operations from its Keyser Valley yards in 1956, the L&WV joined. Rates were published from various stations on the Laurel Line to points on the DL&W, Nickel Plate, Wabash, Monon and other lines. A Scranton truck company, William Land, Inc., was contracted to pick up and deliver trailers from the DL&W ramp to L&WV stations. As originating or receiving carrier, the Laurel Line received $35 per trailer, but it had to pay all expenses from this sum.

During the first month of operation, in May 1956 only 13 trailers moved from Laurel Line destinations. It is doubtful if any new traffic was generated from this service, but Watters was anxious to please the DL&W, which was promoting piggyback shipments in its territory.

All types of traffic was solicited by the L&WV, but the results were generally disappointing. Carloads hauled fell from 7,027 in 1954 to 6,956 in 1955 and to less than 5,000 in 1956. An analysis of freight received at Wilkes-Barre in 1959 showed that, of 1,200 cars, 500 were of fruits and vegetables consigned to wholesaler Thomas C. Thomas Co.; almost 600 were meats consigned to Wilson and Company, Swift and Company or the Lehigh Beef Company; and the remainder were items such as furniture, toys, paper and miscellaneous commodities. A small amount of on-line traffic, consisting of hauling coal from a mine to a washery, ceased in 1955.

A DL&W study in 1956 revealed interchange traffic was divided as follows: 51 percent with the DL&W, 28 percent with the Erie, and 21 percent with the Lehigh Valley. Seventy percent of the traffic consisted of manufactured goods or food products; the remainder was cinders, used in making cinder-block. About 62 percent of the cars originated or ended in Scranton; the rest was handled at Wilkes-Barre.

Watters tried as best he could to cut expenses as business declined. The number of employees dropped from 150 in 1945 to only 34 ten years later. He could do nothing about union rules, however, and every L&WV train, no matter how short, carried a four-man crew, which was excessive under the conditions prevalent on the line. Wages by 1960 were about $1.76 an hour for the engineers and conductors, and 10 cents less for brakemen, only slightly above the $1.60 federal minimum.

Fire and liability insurance coverage was reduced as the plant was dismantled, falling from about $668,000 in 1950 to $136,000 in 1958, thus effecting some savings in premiums. Occasional mishaps added to Watters' burden, such as repairing a bridge over the tracks in Plains, damaged when the driver of a truck hauling about six tons of coal ignored the four-ton load limit posted on the structure, and rerailing a locomotive in Scranton yard and paying for the damage.

Still, the bottom line was depressing. In 1954, 1955 and 1956, the road had operating revenues of about $200,000 annually, but operating expenses averaged about $250,000 each year. When taxes and interest were added, the annual deficit was more than $200,000. It was clear that ending passenger service had not helped the balance sheet in any appreciable fashion. Only Watters' strategy of aligning the L&WV with its large northern neighbor would prevent rapid liquidation of the enterprise.

The engineer of leased Erie-Lackawanna diesel 357 has an unexcelled view of the Wyoming Valley as his northbound freight leisurely crosses Prospect Viaduct just north of Wilkes-Barre's city line in 1961. This trackage was abandoned after the formation of Conrail in 1976 but the viaduct was not removed until the summer of 1985. The former Wilkes-Barre Connecting Railroad bridge over the Susquehanna River, in the background, remains busy today with Delaware and Hudson Railway trains running between New England and Virginia. *Edward S. Miller*

Merger into the DL&W. As the Laurel Line struggled to hold on in the 1950s, the Delaware, Lackawanna and Western Railroad moved to acquire the former electric line. From the point of view of the Lackawanna, the Laurel Line would provide direct access to Wilkes-Barre, something it had never managed to achieve on its own. The ex-electric also served industries and businesses the DL&W could not reach directly, and it provided a limited number of carloads each year to the big carrier's traffic. And the possibility that some rival, such as the Lehigh Valley, might gain control of the Laurel Line could not be ignored. Under Watters' management, a close working relationship had developed between the two lines, as evidenced by the interchange percentages, the piggyback agreement, and the forebearance of the steam road in not pressing for immediate payment of overdue bills.

Quietly, the DL&W began to buy the L&WV's 5 percent, first-mortgage gold bonds, at bargain basement prices. By 1956, the road had acquired about 75 percent of the outstanding bonds, whose owners were no doubt glad to be rid of them. On August 27, the Lackawanna filed a reorganization plan and a request to take control of the Laurel Line. With court approval, the proposal was submitted to the Interstate Commerce Commission, which conducted hearings in February 1957.

In its presentation, the DL&W pointed to the rundown condition of the L&WV as a result of deferred maintenance. Rail replacements were made from steel salvaged from the former southbound main; ordinary track ties were placed on bridges instead of specially treated timber; and the failure to do drainage work had caused a deterioration of the roadbed. The DL&W estimated it would cost about $600,000 to fully rehabilitate the Laurel Line, and about $50,000 a year for normal maintenance thereafter. If the rebuilding were not undertaken, at least $65,000 in maintenance would be required.

In examining the L&WV's financial condition, the Lackawanna noted it had obligations of over $2,900,000, with an additional burden of $2 million in unpaid interest. On paper, the property had a value of about $4,400,000, but its scrap value was put at only $200,000. The DL&W petition concluded the line could never earn its way back to solvency with its current overvalued capitalization.

If its plan won approval, the Lackawanna estimated the Laurel Line could earn a profit of about $8,000 a year. What was the basis for this optimism about a carrier that was currently posting a deficit of $60,000 in operating revenues alone? The DL&W projected savings of $80,000 by closing the Laurel Line yard, freight station and offices and running the line as a branch with its regular equipment and personnel. It also expected an increase in traffic based on normal growth, the new piggyback service, and the location of new industries along the tracks of the former electric road. This latter development could only occur if shippers were confident the Laurel Line would continue to run, which was not likely unless the DL&W assumed control.

The DL&W financial plan was simple. The new company, to be incorporated as the Lackawanna and Wyoming Valley Railway Company (rather than Railroad Company) could issue a total of $600,000 in Prior Lien Mortgage Bonds (to finance a rehabilitation program). The regular capitalization would consist of $975,000 in General Mortgage, 4 percent, 50-year income bonds, and $975,000 in common stock. These would be issued to the present bondholders on the basis of a $500 par value bond and a $500 par value share of common for each $1,000 First Mortgage bond. Interest and dividends would be paid, however, only if there were sufficient net earnings to cover them. The old 6 percent debentures, and the preferred and common stock would have no value and would be discarded. Since the DL&W held most of the old First Mortgage bonds, it would be the prime owner of the new railway. But, as the petition concluded, "Simply expressed, reorganization is essential to continued service, and continued service is in the public interest."

Although the hoped-for new industries would not appear, and consequently the rehabilitation would never occur, the plan was a sensible one and offered the only prospect of continuing operations along the route. The stock and bondholders had been aware of the bleak financial outlook for more than 25 years and the write-off of their investment, while unpleasant, could hardly have been a surprise. The line had been heavily overcapitalized from its inception and it had only periodically earned a profit, and never one sufficient to pay off its debts.

While, to a degree, the Lackawanna plan was self-serving, it would also benefit the community by continuing operations. The fact that it was based on an overly optimistic outlook regarding new industry in the region and therefore failed should not be held against it.

The court accepted the DL&W plan and the ICC gave its approval on December 6, 1957. A new trustee was given control of the property and on August 19, 1959, the Lackawanna and Wyoming Valley Railway Company was incorporated. This firm, which was a creature of the DL&W, officially took over on February 24, 1960, but by then the line had been operating for several years as a Lackawanna Railroad branch line.

Thus ends the story of the Laurel Line as an independent corporation. Conceived in an era of unrestrained free enterprise and based on grandiose plans, it nonetheless fulfilled a more mundane but basic purpose of serving the people of the valley region for more than half a century.

Although the merger between the Erie and the DL&W had been consummated, diesel 357 (RIGHT) although renumbered, was still lettered "Lackawanna" as it pulled an L&WV freight south on Sunday, May 13, 1962. The train is running on the former Erie Railroad tracks at Moosic, directly above the idled former L&WV rails. The diversion of traffic off the old L&WV, owned by the DL&W, resulted from the formation of the Erie-Lackawanna. With the corporation now owning two railroad lines in the area, the least likely to earn a profit was closed. The small Brady Street yard (BELOW) was installed adjacent to the abandoned Scranton car house in February, 1968, as part of the Spruce Street Bridge highway project. Although intended as an interchange point between the L&WV and the Erie-Lackawanna, L&WV crews continued to pick up their train at the E-L yard. The train shown had 18 cars trailing the diesel. *BOTH: Edward S. Miller*

Postscript: The Laurel Line Today. Although it is not within the scope of this book to examine the affairs of the L&WV as an insigificant branch of much larger rail systems, a brief word is in order to describe the line's gradual dismantlement. By the time the DL&W was completing its takeover of the little ex-electric road, a much larger merger was being effected which would result in the abandonment of almost half of the L&WV branch.

By the late 1950s, the railroad industry as a whole was suffering from many ills which threatened the stability, and even the existence, of some carriers. The DL&W, which extended from Hoboken, New Jersey, to Buffalo, New York, was beset with competition from all modes of transportation and, like other carriers, looked to merger as a way of easing its difficulties. For various reasons, the line began discussions with the Erie, which culminated on October 17, 1960, with the formation of the Erie-Lackawanna Railroad Company (E-L).

For a short period after the merger, the L&WV branch operated as it had since the mid-fifties, with several trains a week serving the on-line industries. Since one purpose of the merger was to eliminate duplicate facilities, the new management decided to combine part of the L&WV with the Wyoming branch of the former Erie, which closely paralleled the ex-electric between Virginia, immediately south of Scranton, and Inkerman, between Pittston and Plains. In August and September 1961, track crews installed about 1,500 feet of trackage at Virginia to connect the Erie line with that of the L&WV. At Inkerman, an existing Erie siding was extended about 500 feet and linked with the Laurel Line track.

On September 14, trains on the L&WV branch began using the new route and nine miles of the old electric line were abandoned. By the next year, work was underway removing the rails, ties, the Pittston station, the Avoca substation and numerous bridges, although the spectacular 682-foot viaduct at Avoca was not dismantled until 1965.

Other changes made by the E-L included rebuilding the track between Connell Junction, about one-half mile south of the tunnel, and Virginia, to permit the use of heavier locomotives. Even tie plates were installed, a luxury for the Laurel Line. Beginning in October 1961, the L&WV freight began hauling a caboose, usually Lackawanna 818, despite the fact that the old L&WV had managed to do without one during its entire existence. In electric days, the crew rode the locomo-

Although one would hardly know it, this photograph shows the former site of Laurel Junction (LEFT), along Roaring Brook, opposite Nay Aug Park, Scranton, as it appeared on Sunday, April 5, 1964. The view to the south shows preliminary grading work underway for future Interstate highway 81. The remains of the ore roaster are still visible in the center of the photo. Hardly recognizable in this scene taken in October, 1983, is the former car house (LOWER LEFT) near Harrison Avenue, Scranton. The building, considerably modified, is now a plastics factory owned by Poli-Hi-Menasha Corporation. Trackage in Brady Yard, in the foreground, is now owned by Conrail. Tracks beyond this point are owned by Pocono Northeast Railway. Heavy cranes labor to undo the work done by the King Bridge Company of Cleveland in 1903, as they remove the span over South Main and Oak Streets in Pittston (LOWER RIGHT) on May 27, 1962. The bridge on the left carries the former Erie Railroad tracks. *ALL: Edward S. Miller*

tive, but that was no longer deemed proper under the new management. Otherwise, the routine continued as before. When needed, the E-L diesel would leave Scranton about 6:00 A.M. and Wilkes-Barre about 9:00 A.M. serving the appropriate sidings and running up the Minooka branch as needed. Gradually the business on the Wilkes-Barre end, always the weaker of the two, failed, as the meat-packing houses curtailed their operations. Most of the remaining business was devastated by the effects of Hurricane Agnes, which flooded part of the city in June 1972.

By the 1970s, the entire Northeast railroad scene was thrown into chaos following the collapse of the Penn-Central Transportation Company on June 22, 1970. Line after line plunged into bankruptcy, including the Erie-Lackawanna itself. Out of the wreckage emerged the federally supported Consolidated Rail Corporation in 1976, which united the bankrupts into one gigantic corporation. In the Lackawanna/Luzerne county area, only the Delaware and Hudson Railway managed to cling to a precarious independence. The creation of Conrail would not only bring about tremendous changes among the former private carriers but it would also affect even the tiny L&WV branch.

At first, Conrail management was too preoccupied trying to organize its rundown properties to pay much attention to the Lackawanna Valley, but soon the Philadelphia-based executives looked northeast. With direct access to Wilkes-Barre available on the ex-Lehigh Valley, there was no need to maintain the former Laurel Line route into the city. By the end of 1976, Conrail had ceased to run trains south of Virginia, although for a period the tunnel was still used to reach the Minooka freight line. In 1980, however, management decided to utilize the former Erie line for this purpose, and the tunnel fell into disuse.

Ironically, in 1983 one of the largest segments of the old Laurel Line still intact was the newest and most poorly constructed section, the three-mile Minooka branch. Technically, this line was never owned by the L&WV, since title rested with the Scranton Industrial Development Company and the General Electric Company. No services were offered in 1983 as the factories located at the industrial park do not depend on rail transportation. It is unlikely trains will ever climb the rugged hillside again.

The only portion of the original road in service at this writing was the short stretch from the former DL&W interchange at Cedar Avenue, through the site of the Scranton terminal, to a small yard opposite the old carhouse, now a factory. Much of the track on the remaining segments of the line, between Scranton and Virginia, and between Inkerman and Wilkes-Barre, was still in place, although considerable portions on the southern end have been removed and the roadbed dug up to recover the culm on which it rests, which is still a marketable commodity when it can be cheaply obtained.

As of 1985, the rail scene in the Lackawanna Valley was in a state of flux. A new short line, Pocono Northeast Railway, which operates various branches cast off by Conrail, acquired the Laurel Line trackage from Virginia, through the tunnel to Scranton. There was talk of running passenger trains from Wilkes-Barre to Scranton, possibly utilizing this segment. Some promoters suggested running a trolley line through the tunnel and building an extension to serve a ski resort at Montage, south of the city. Great hopes for an economic revival of Scranton were placed on Steamtown, a rail museum which announced it would move from Bellows Falls, Vermont, to the former DL&W yards and shops in the downtown section of the city. Whether these developments materialize, only time will tell.

Not surprisingly, for a line so well constructed as the L&WV was, there are many traces of it still in existence. The visitor who so desires may follow much of the route of the old Laurel Line, even though some portions have been obliterated by highway construction. Whatever the future may hold, the memory of the Laurel Line will long linger among the hills and valleys of Lackawanna and Luzerne counties.

While workers look from above, the fourth span from the Pittston end of the great Avoca Viaduct falls into the gorge at 3:50 p.m. on Thursday, November 11, 1965. The dismantling began the week of October 17, 1965 and was not completed until March 19, 1968.
Edward S. Miller

8

Conclusions

THE LACKAWANNA and Wyoming Valley Railroad lasted less than 60 years as an independent corporation. Conceived at the beginning of the 20th century in a time of unbridled optimism about the prospects of electric railways, it eventually was merged into a large steam railroad and was gradually dismembered. What can be learned from its half century of operations? Was the line justified in terms of service, function and economics? How well was it managed? What impact did it have on its surroundings? In short, was it worthwhile? While no absolute answers can be made to these questions, the authors will attempt to offer some possible conclusions.

The basic idea of a high-speed electric railway, linking the twin anthracite capitals of Lackawanna and Luzerne counties, was a sound one. Local lines could not meet the need, and the major steam roads would not. However, there were a number of flaws in the initial proposal.

The promoters, as was usually the case, painted an overly bright picture of traffic and profits. As originally planned, the line was to extend, eventually, from Hazleton to Carbondale and beyond, but only the Wilkes–Barre–Scranton line and what became the short branch to Dunmore were actually constructed. The main line was extremely well-built, too well-constructed as it developed, and the Dunmore branch was a weak spur that never justified its existence once the possibilities of a northern extension had been extinguished.

A basic misconception was the hope that freight traffic could be lured onto the line, turning it into a lucrative bridge road, linking the manufacturing and agrarian lands to the south and west with the New York–New England market. It was totally unrealistic to expect the major railroads to divert traffic from their own tracks to that of a fledgling shortline, and one operated by electricity at that. The anthracite region was overbuilt with railroads, and while they may have been slow and cumbersome, they did provide many alternate routes, which were already well established.

Generally, too, they were hostile to electric lines as a potential threat, making the likelihood of joint service extremely remote. The early promoters and managers seemed unaware of this, and for many years maintained a rosy outlook about the prospects for through freight traffic.

If the promoters had been more realistic, they would have concentrated on building a good railroad, but not one to the L&WV's high standards. With its many bridges and structures, its fine double-track, its private right-of-way with only a few grade crossings, its impressive powerhouse, carbarn and stations, and soon its Scranton tunnel, the Laurel Line was overbuilt. It was saddled with a tremendous debt before it ever turned a wheel, and during its entire life this bonded indebtedness hung like the fabled albatross around its corporate neck.

A more modest physical plant would have been justified, one designed for passenger trains but not necessarily for heavy freight traffic. The company's late arrival on the scene forced it to skirt the edges of all its intermediate towns, except Pittston, and the expensive Scranton tunnel, which provided an excellent entrance to downtown, had far more value for passenger service than for freight.

Why did the promoters, and Westinghouse, overbuild? From the perspective of 1984, it is impossible to say. A common belief, often expressed among the railfan fraternity, was that Westinghouse wanted the line "for experimental purposes" and as a demonstration of the quality of Westinghouse equipment. The motives of the Westinghouse group are unknown, but it seems likely that they regarded the L&WV as a good investment, which would earn a profit and also which would serve as a working example of the company's products. There is no evidence that the Laurel Line was ever used in an experimental way.

Although it did belatedly purchase the 1895 pioneer locomotive, the machine had been set aside by its builders and one cannot avoid the impression that Westinghouse was glad to unload it at a good price on its unlucky subsidiary. Further, the L&WV was a more or less standard 650-volt, third-rail, DC electric road, with little of the daring or innovative in its electric structures. George Westinghouse himself, busy with his experiments on AC electrification which would soon bear fruit on the New York, New Haven and Hartford

Railroad, seemed hardly aware of the existence of the anthracite region carrier.

It seems likely that a few of the promoters syphoned off some of the line's funds. Records were not kept; the contractors operated the line initially; a general carelessness seems to have prevailed. The total cost of more than $9 million for a 20-mile railroad places the L&WV as one of the more expensive carriers built. Promoters in the age of unrestrained and unregulated capitalism often found ways to take their profits "from the top," and although it cannot be proven, the authors feel the Laurel Line lost about a million dollars to the pockets of some of the promoters and contractors. If this conjecture is correct, the L&WV was forced to carry an additional, unnecessary burden.

As has been demonstrated, the line was not particularly well managed during its corporate lifetime. American business has frequently practiced a "hit-or-miss" style of control, with executives assuming responsibilities and often relying upon "on-the-job training" for their expertise. In this respect, the L&WV was probably no better nor worse than its contemporary railway companies and, once operations were underway, the quality of management was generally adequate, but certainly not outstanding or spectacular.

The first of the three phases of control saw the flamboyant and ever-optimistic Charles Conn exercising authority in the name of the Westinghouse group. Neither very experienced nor particularly gifted, Conn often seemed more impressed with his position, his title and his standing in the community than he did with his obligation to establish a profitable and soundly run company. Nor was he aided by the absentee control of the Westinghouse organization, personified by President George Smith in New York. Conn's experiences in developing on-line parks were not particularly fruitful and his pursuit of through freight arrangements with the connecting steam roads can only be judged a failure. Eventually, Westinghouse became disillusioned and sold out at a loss.

The Sproul syndicate ownership marked the second phase of Laurel Line management. Initially, the new owners, recognizing their inexperience, relied on the engineering firm of Ford, Bacon and Davis to guide the line. While Frank Ford especially showed considerable interest in the road, he had many other obligations, and only occasionally was he on hand in Scranton to make decisions.

Fortunately, he had an able ally in Patrick J. Murphy, who gradually assumed complete control as the financial picture darkened and Ford faded from the scene. While sometimes petty and smug in his approach, Murphy was conscientious, diligent and involved on a daily basis with the line's affairs. He was closer to being an "operating man" than the road's other managers. After the sobering experience of the 1920 wreck, Murphy seemed to settle down and grew with the job. He certainly must be credited with the will to keep the line running during the bitter Depression years.

By the time the last of the trio of managers took over, the Laurel Line was struggling for survival. E. McLain Watters was particularly well suited to the task of trying to unravel the financial knots and save the line from outright abandonment. He early recognized that merger was the only road to salvation and, even though he was forced to discontinue passenger service and electric motive power, his administration must be judged competent. It is doubtful if anyone could have done more.

The forces of organized labor also performed reasonably well from an overall standpoint. Electric railroads were the "glamor industry" of the day when the road opened, and the company was fortunate in attracting many qualified young men who remained with it for its entire life. There were difficult times during the hectic World War I era, and certainly carelessness contributed to the fatal 1920 wreck.

Considering their wages were relatively low, and often inferior to those of neighboring steam roads, the workers of the Laurel Line performed their daily tasks with a general air of competence and skill. As decline set in following World War II, the inflexibility of rail labor in insisting on full crews and annual raises did not help, but it is unlikely that any concessions on the part of labor would have appreciably changed the line's fate.

More so than most railroads the Laurel Line was heavily dependent on the economy of the region it served. Since Lackawanna and Luzerne counties were dependent in large measure on the anthracite trade, and since the Laurel Line was so short, it was drastically affected by the fall of anthracite mining, especially after the 1920s. Strikes, closures, mine failures, the loss of markets and falling population counts impacted heavily on the L&WV, severely curtailing both its passenger and freight revenues.

Whether the line could have survived even in the present age of public subsidies and ownership is highly problematical, for the valley residents seem closely wedded to their automobiles and the central business districts of Scranton and Wilkes-Barre are in all likelihood too weak to support a rapid transit operation such as the Laurel Line offered.

In summary, the L&WV met a definite transportation need, especially of people, until the 1940s. Its technology was sound if not advanced, and it gave fast, frequent service to the valley region, carrying prodigious numbers of passengers. In time, its crushing debt and falling traffic base, both passenger and freight, led to its acquisition by its large neighbor. Abandoned in piecemeal fashion, parts of it still remain, mostly idle, but perhaps the final chapter has yet to be written along the route of the Lackawanna and Wyoming Valley Railroad.

As combine 114 enters the Scranton terminal in 1935, the young conductor stands on the step ready to drop off as the car stops. *George Krambles*

Roster of Equipment

PASSENGER COACHES

CAR NO.	YEAR	BODY TYPE	WEIGHT	COST	BUILDER	MOTORS	CONTROL	DIMENSIONS	TRUCKS	SEATS	NOTES
1-10	1902-03	wood Type L	72,000 lbs.	16,000	ACF Co. Jackson & Sharp Plant	2 WH 50F; 150 HP gear ratio: 19-52	L-2 drum control; later WH DE-AB	length: 52' width: 9' height: 13'	Baldwin MCB wheel base 6' motor tk: 36" trailer: 33" spoke wheels	61: reblt: 64	1, 29
11-23 10(II)	1904	wood; steel underframe; Type M	78,000 lbs.	13,000	John Stephenson Car Co.	2 WH 50F: 150 HP gear ratio: 18-59 replaced: 1919-20 WH 557-A8 motors 140 HP	Turret type unit switch group; replaced by WH DE-AB	length: 51'2" width: 8'10" height: 13'	Baldwin MCB wheel base 6' motor tk: 36" trailer: 33" spoke wheels	64	2,29,30
31-34	1924	steel	80,000 lbs.	27,000	Osgood Bradley Car Co.	2 WH 557-A8: 140 HP gear ratio: 18-59	WH type DE-AB Tomlinson couplers	length: 60'6" width: 9' height: 13'2"	Baldwin MCB wheel base 7' motor tk: 36" trailer: 33"	72	3,29
35-36	1925	steel	80,000 lbs.	27,000	Osgood Bradley Car Co.	2 WH 557-A8; 140 HP gear ratio: 18-59	WH type DE-AB Tomlinson couplers	length: 60'6" width: 9' height: 13'2"	Baldwin MCB wheel base 7' motor tk: 36" trailer: 33"	72	3,29
37-39	1928	steel	84,500 lbs.	21,000	Osgood Bradley Car Co.	2 WH 557-A8; 140 HP gear ratio: 18-59	WH type DE-AB Tomlinson couplers	length: 63'2" width: 9' height: 13'2"	Baldwin MCB wheel base 7' motor tk: 36" trailer: 33"	76	3,4,5,29

COMBINATION CARS

CAR NO.	YEAR	BODY TYPE	WEIGHT	COST	BUILDER	MOTORS	CONTROL	DIMENSIONS	TRUCKS	SEATS	NOTES
101-104	1903	wood Type L	72,000	16,000	ACF Co. Jackson & Sharp Plant	2 WH 50F; 150 HP gear ratio: 19-52	L-2 drum control; later WH DE-AB	length: 52' width: 9' height: 13'	Baldwin MCB wheel base 6' motor tk: 36" trailer: 33" spoke wheels	44	6,7
111-113	1904	wood; steel underframe; Type M	78,000	13,000	John Stephenson Car Co. replaced: 1919-20	2 WH 50F; 150 HP gear ratio: 18-59 WH DE-AB WH 557-A8 motors: 140 HP	Turret type unit switch group; replaced by WH DE-AB	length: 51'2" width: 8'10" height: 13' trailer: 33"	Baldwin MCB wheel base 6' motor tk: 36" spoke wheels	44	7,8,29
114-115	1925	steel	80,000	27,000	Osgood Bradley Car Co.	2 WH 557-A8; 140 HP gear ratio: 18-59	WH type DE-AB Tomlinson couplers	length: 60'6" width: 9' height: 13'2"	Baldwin MCB wheel base 7' motor tk: 36" trailer: 33"	48	3,9,29
116	1928	steel	84,500	21,000	Osgood Bradley Car Co.	2 WH 557-A8; 140 HP gear ratio: 18-59	WH type DE-AB	length: 63'2" width: 9' height: 13'2"	Baldwin MCB wheel base 7' motor tk: 36" trailer: 33"	52	3,4,9,29

FOOTNOTES ON PAGE 203

The platform side of type L coach number 1 was recorded as it rested on shop trucks outside the Jackson & Sharp plant in Wilmington, Delaware, late in 1903. Destined to run in this configuration for only three years, the car was extensively remanufactured as a type M coach in 1907. Not only did this require extensive electrical and mechanical modifications cited elsewhere in this book, but it also required that the interiors of the cars be gutted. The original forward-facing seats were discarded, after the Laurel Line was unsuccessful in an attempt to return them to the builder for credit, and replaced with Hale & Kilburn model 99F *Walkover* seats, enabling passengers to face forward regardless of which direction the cars operated. The car ends were stripped and replaced. Steps, traps and doors, as well as all door and trap hardware were also replaced. The 1906 minute books show that Westinghouse Traction Brake Co. supplied at least one new D-2EG compressor, replacing the original air pump on the first converted car. Essentially, only the trucks, motors, and the central portion of the body were retained. An entry in the 1907 minute books shows a cost to convert 14 cars of $138,806.51, compared to the $224,000 that had been expended only four years previously to buy them. The scale drawing by Charles R. Yungkurth shows the original blind side configuration. *Photo: George Krambles*

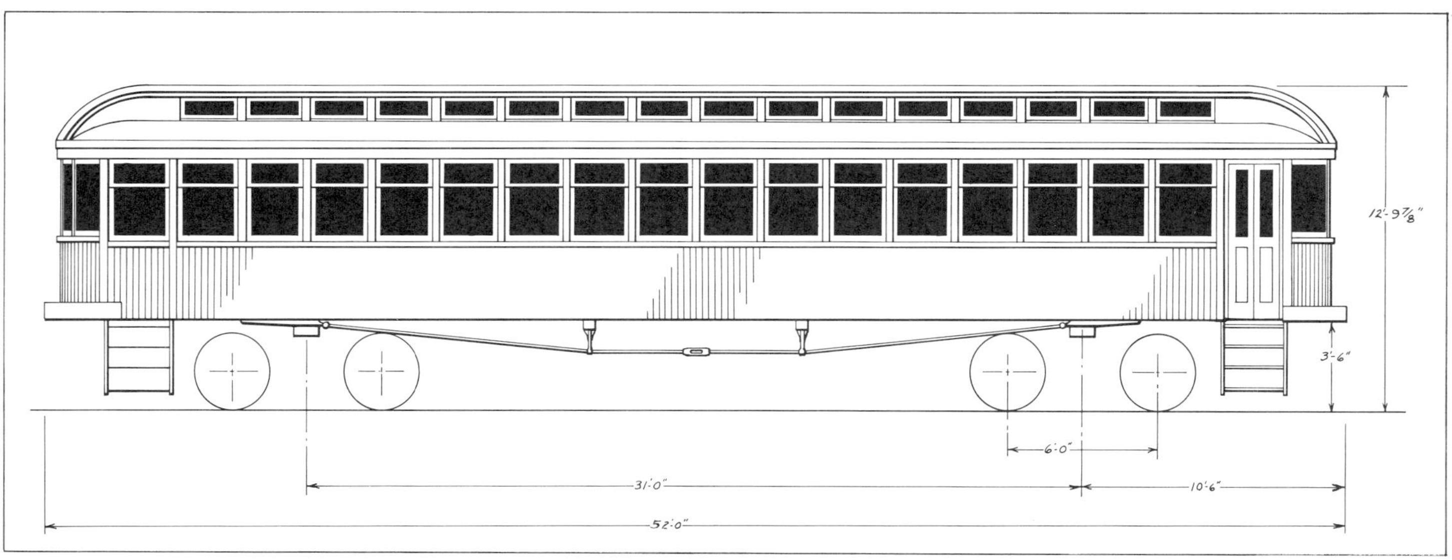

Type L combination car 101 was posed at Scranton in 1904 for this equipment record photo. *Pennsylvania Historical & Museum Commission* American Car & Foundry drawing E-128 appears to be the left side of this same type car. It may differ, however, in minor details from the Laurel Line equipment. *Edward S. Miller Collection*

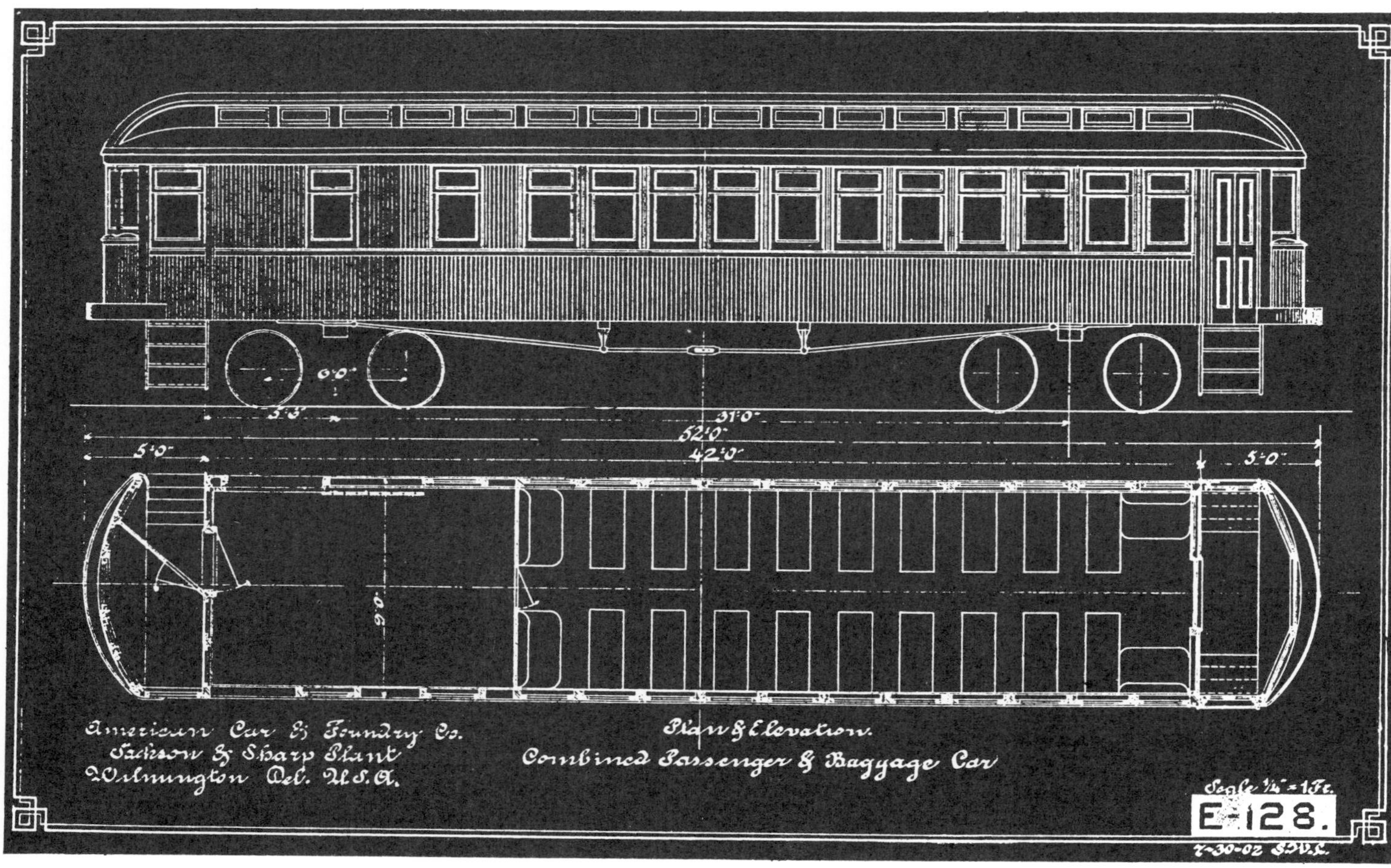

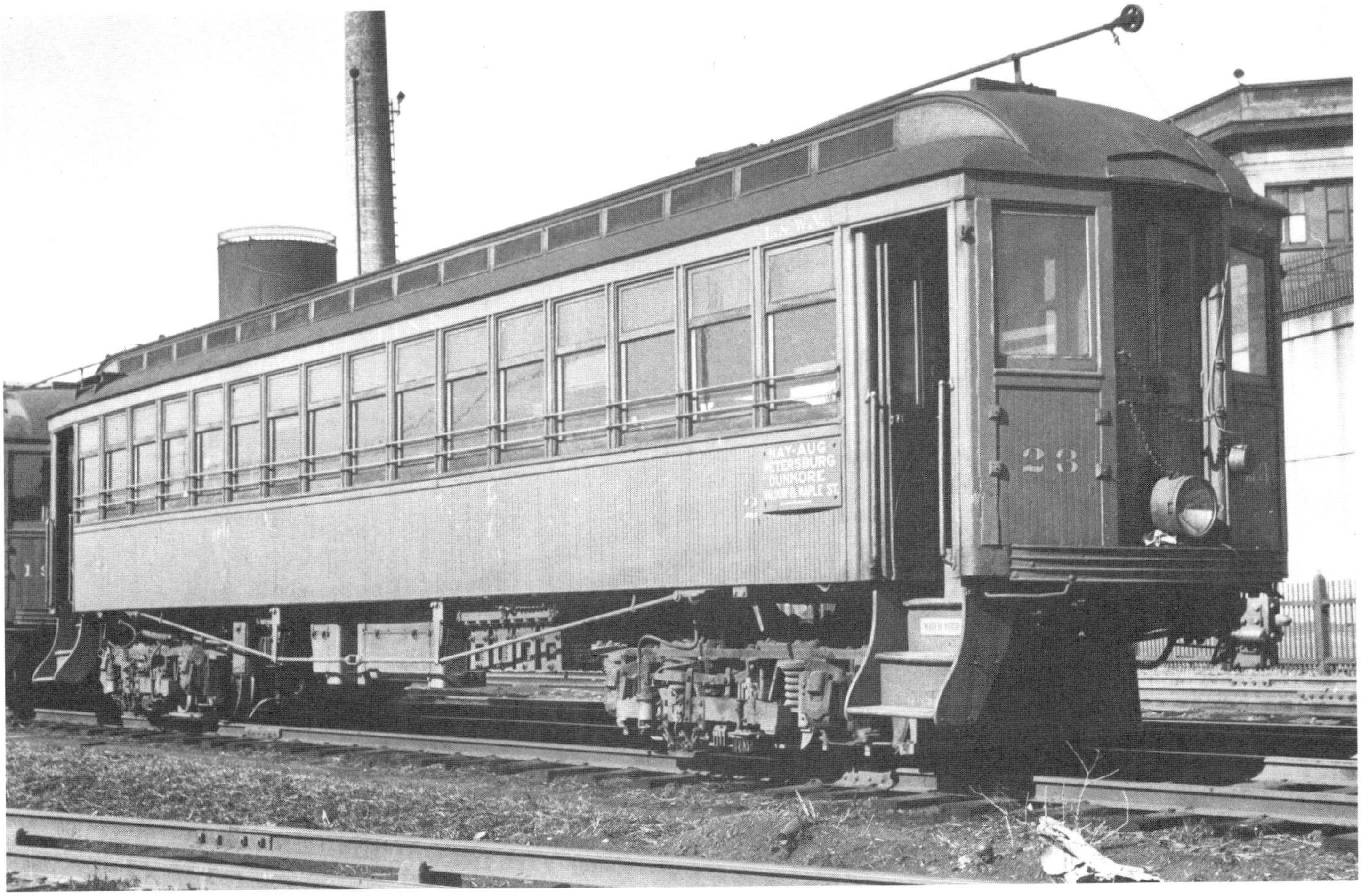

The right side of type M car 23, (ABOVE) shown rebuilt for one-man service on the Dunmore branch, displays the underfloor resistor grids, switch group, reverser, and line breaker cases, and the battery box. All of the propulsion hardware was on this side of the car. The air compressor was at the rear of the other side. The air tanks were mounted under the middle of the car, requiring a little stretching to reach the water drain cocks for daily servicing. The steps on the motorman's side of the front had been removed. *Michael J. Lavelle, Edward S. Miller Collection* **The left or compressor side of type M car 22 is displayed in Wilkes-Barre in 1935. This car, between 1938 and 1942, was the only remaining type M car in the two-man configuration.** *James P. Shuman*

The interior of the type M cars (LEFT) was quite similar to the rebuilt type L equipment, right down to the red plush upholstery. Above the door to the smoking section is the 1903 prohibition against expectorating which carried a stiff fine of $1.00. No doubt, when this picture was taken in 1945, the fine was less a deterrent than common sense. *Edward S. Miller*

Type M combine 113 (BELOW) rests between assignments at the excursion platform in Scranton during 1942. *Edward S. Miller* **Originally, these cars were fitted with Van Dorn couplers (RIGHT). All control functions were passed from car to car through jumper cables. These cables were plugged into receptacles under the buffer. The two heavier air hoses connected the main air storage reservoirs and the brake pipe on adjacent cars. The lighter hose was used for communicating signals. Later Tomlinson couplers with attached air and electric portions were installed; when coupling the cars thus equipped, air and electric connections were made automatically.** *Charles Sobeck, Edward S. Miller Collection*

Car 34 (ABOVE), at Scranton in 1941, was typical of the short version of the standard Laurel Line steel coach. The wider window post marks the interior bulkhead dividing the smoking and non-smoking compartments. The car is painted in the tuscan red paint scheme. Car 32 (BELOW), from the same order, models the flamboyant vermilion and blue livery applied first to this car in 1949. The use of the blue paint was downplayed on subsequent cars. *BOTH: Edward S. Miller*

Varnished wood in conjunction with dark plush or leather upholstery made the interiors of the steel cars sombre. Car 39 (UPPER RIGHT) was typical. *William C. Janssen* **Brighter but the subject of many unflattering remarks by conservative patrons was the painted interior of car 32 (UPPER LEFT). Two shades of gray, white and cream paint were applied in 1949 in a never repeated experiment.** *Edward S. Miller*

Car 116 (LEFT) was the longest baggage-passenger combine on the railroad. Note the Hyatt roller bearing journals on the front truck. Plain bearing journals were applied whenever the roller bearings became unserviceable. The illustration was taken in Scranton in 1939. A truck with plain bearings (ABOVE) was used under combine 114 at the same time.
BOTH: Edward S. Miller

EXPRESS CARS

CAR NO.	YEAR	BODY TYPE	WEIGHT	COST	BUILDER	MOTORS	CONTROL	DIMENSIONS	TRUCKS	NOTES
201-204	1903	wood	72,400	9,000	ACF Co. Jackson & Sharp Plant lot #2431	2 WH 50F; 150 HP gear ratio: 19-52	L-2 drum control	length: 52'2" width: 9'8" height: 13'8"	Baldwin MCB wheel base 6' motor tk: 36" trailer: 33" spoke wheels	10

The express cars retained many features that had been removed from the type L passenger cars in 1907, such as the five-window front end, the drum controls, and even the sand tank mounted on the front truck. Three photos are required to illustrate that these cars were totally lacking in symmetry. The two photos of 202 were taken in 1939 (RIGHT) and 1947 (ABOVE). The rear view of derelict 204 dates to 1951. *RIGHT: Edward S. Miller DRAWING: Charles R. Yungkurth; OTHERS: Frank L. Watson from Richard S. Short Collection*

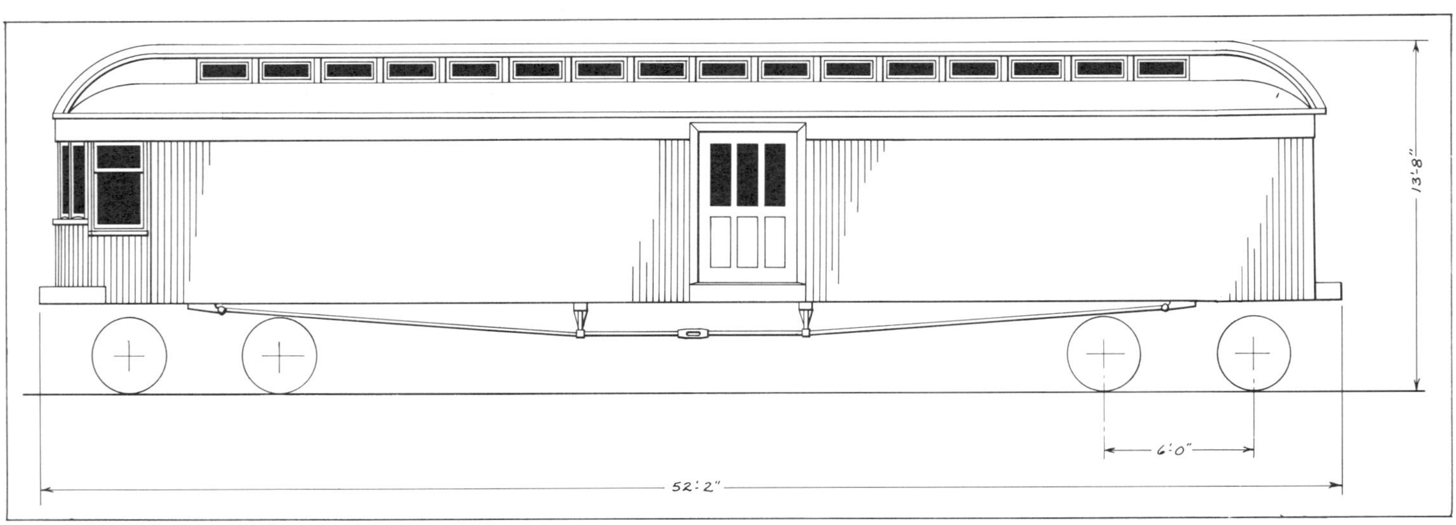

ELECTRIC LOCOMOTIVES

CAR NO.	YEAR	BODY TYPE	WEIGHT	COST	BUILDER	MOTORS	CONTROL	DIMENSIONS	TRUCKS	NOTES
401	1895; acquired 1906	box cab	97,000	17,500	Baldwin-Westinghouse	4 WH motors; 150 HP; rebuilt 1923; 4 WH 577-A6 195 HP	WH unit switch control, HL	length: 32' width: 9' height: 12'5"	Baldwin wheel base 7'6" 36" wheels	15
402	1907	box cab	106,000	17,500	Baldwin-Westinghouse	4 WH motors; 150 HP; rebuilt 1923: 4 WH 577-A6 195 HP	WH unit switch control, HL	length: 32' width: 9' height: 12'5"	Baldwin wheel base 7'6" 36" wheels	16
403	1917-1919	box cab	106,000	20,000	Scranton shops, L&WV	4 WH 557-A8; 140 HP (see note)	WH unit switch control, HL	length: 32' width: 9' height: 12'5"	Baldwin wheel base 6' 36" wheels	17

Two photos show 401 as it appeared during the Westinghouse alternating current experiments after 1895 *(ABOVE by Charles A. Brown from George Krambles)* **and freshly painted on the Laurel Line in 1939** *(LEFT by Edward S. Miller).* **The air compressor, air reservoirs, switch group, line breaker, reverser, resistor grids and other mechanical hardware were mounted inside the body. Credit Charles R. Yungkurth for the drawing of the same engine.**

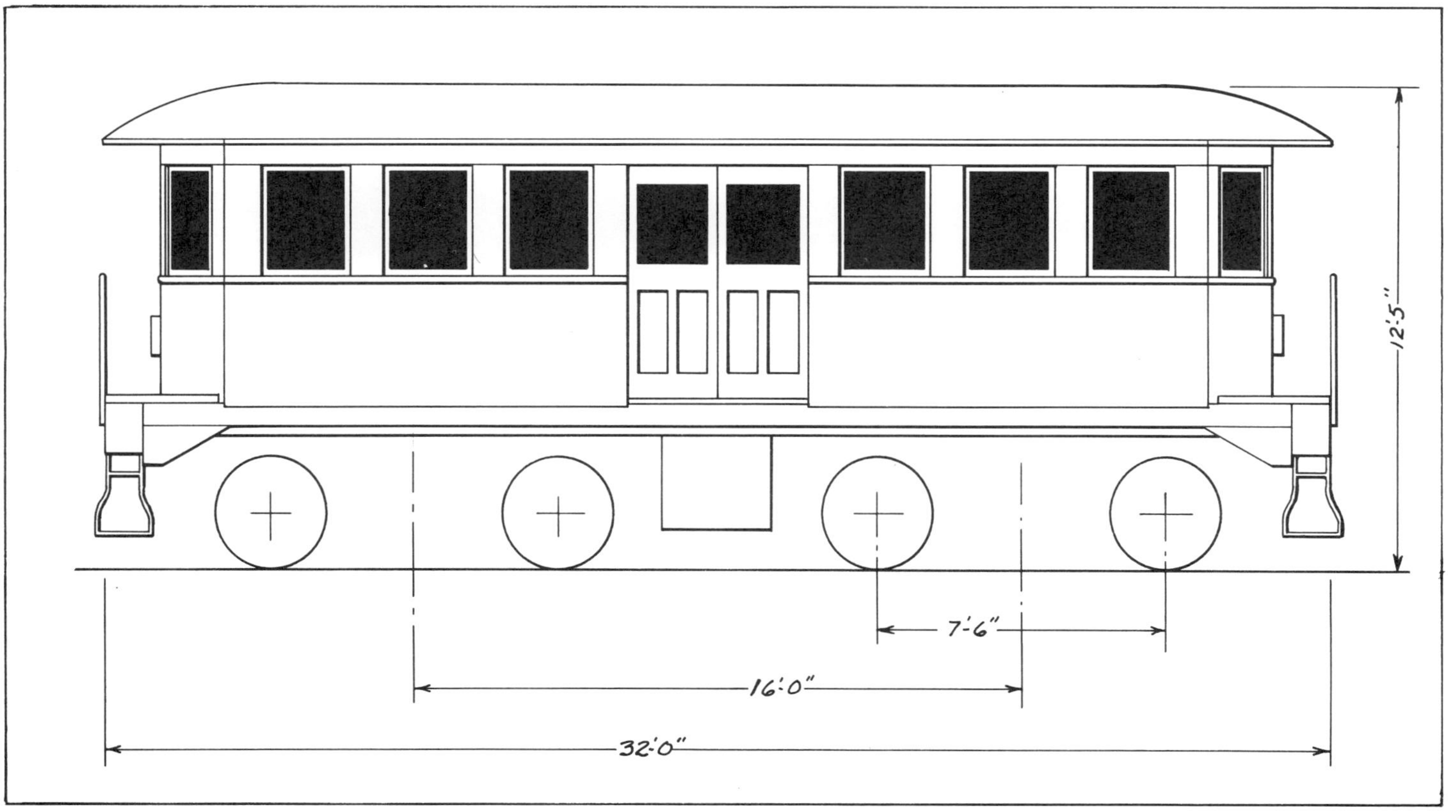

FREIGHT CARS

NO.	YEAR	TYPE	WEIGHT	COST	BUILDER	DIMENSIONS	NOTES
500-509	1903	box	32,700 lbs.	9,900	American Car & Foundry Berwick, Pa.	30 ton capacity	20
601-605	1909	flat gondola	28,000 lbs.	2,500	Hicks Locomotive Works	length 36'; sides 20" 40,000 lbs. capacity Westinghouse air brakes	21,30
1001-1002	1902	flat				40,000 lbs. capacity	22,30
1003-1005	1905	gondola		400	purchased used	40,000 lbs. capacity	23
701-710	1906	hopper bottom gondola	22,500 lbs.	2,731		50,000 lbs. capacity	24
711-720	1907	hopper bottom gondola	22,500 lbs.	345 ea.			25
721-740	1917	wooden gondola		3,000			26
751-765	1914	steel hopper		13,800	Pressed Steel Car Co., McKees Rocks, Pa		27,30
711-716(II)	1921	ash					28,30
717-720(II)		ash					
801-802	1906	Rogers Ballast	35,000 lbs.	700		86,000 lbs. capacity	

Four photos from Edward S. Miller show typical L&WV freight cars. Clockwise from upper left are wooden box car 508, hopper-bottom gondola 715, steel hopper car 753, and flat car 1005. The use of arch-bar trucks eventually precluded interchange of these cars with other railroads.

WORK EQUIPMENT

CAR NO.	YEAR	BODY TYPE	WEIGHT	COST	BUILDER	MOTORS	CONTROL	DIMENSIONS	TRUCKS	NOTES
301	1903	Single-end cab flat car	61,300	8,000	Amer. Car & Foundry, Berwick, PA	4 WH #76; 75 HP	L-2 drum control			11
302	1906	work flat car			rebuilt by L&WV					12
303		work flat car with tank			rebuilt by L&WV					13
802	1906	derrick car			rebuilt by L&WV in 1919			length in front of cab: 28' width: 9'6"		14

This potpourri of non-revenue cars includes 502 (ABOVE), a box car used by the line department; 605 (RIGHT), used for weed spraying; and 301 (BELOW), the work locomotive acquired when the line opened in 1903.
ALL: Edward S. Miller

Although unsuited for rerailing errant rolling stock, crane 802 was sufficiently heavy to be useful for track renewal tasks. Here it is being used to help renew crossings at the Miner Hillard spur in Wilkes-Barre in 1945. The lack of paint would indicate that it was used infrequently and received equally sparse maintenance.
Edward S. Miller

STEAM LOCOMOTIVES

CAR NO.	YEAR ACQUIR.	BODY	WEIGHT	COST	BUILDER	CYLINDERS	BOILER PRESSURE	DRIVERS	NOTES
4	1901	4-4-0	73,000 lbs.		Lackawanna & Bloomsburg Railroad, Kingston, Pa., 1875	16"x24"	130 lbs.	63"	18
10	1902	2-6-0	98,000 lbs.	3,600	Dickson Locomotive Works 1873; construction number 113	18"x24"	140 lbs.	57"	19

MOTOR BUSES

NO.	YEAR	BUILDER	BLDR. NO.	ENGINE NO.	MODEL	CYLINDERS	SEATS	DATE OF PURCHASE	DATE OUT OF SERVICE	NOTES
15	1926	Mack	539483	BA560100b	AB	4	25	7/26	5/32	31
17	1926	Mack	593520	BA56060B	AB	4	25	7/26	3/32	
18	1926	Mack	593519	BA56044B	AB	4	25	7/26	2/32	
19	1926	Mack	592211	BA44914B	AB	4	25	4/26	12/33	
20	1930	Mack	6B035102	BCE-10-9	BC	6	29	8/30	11/37	
21	1930	White	168713	360	65	6	25	12/30		
22	1932	White	176736	397	65A	6	29	12/32	9/37	
23	1932	White	176252	280	65A	6	29	2/32	11/37	
24	1932	White	177296	4A-154	613	6	21	5/32	1937	
25	1934	Dodge	8357860	T-69979	K34	6	19	4/34	1936	
26	1934	Dodge	8365481	T-621037	K34	6	19	9/34	9/37	
—	1936	Dodge	8390733	T-26-2361	LF37A		21			
—	1936	Dodge	8395516	T-26-1136	LF37A		21			
—	1937	Ford	417052	4170521	70		27	10/37		
—	1937	Ford	4199086	4199086	70		27	10/37		
32	1937	Ford	3906248	3906248	70		27	10/37		
33	1937	Ford	4196151	4196151	70		27	10/37		
34	1938	Ford	4582998	4582998	81B		27			
35	1940	Ford	109120	109120	09-B		27	12/39		
36	1945	Ford			59-B		27			
37	1946	Ford			69-B		29			
38	1946	Ford			69-B		29			
39	1944	Ford			49-B		29			
40	1944	Ford			49-B		29			
41	1945	Ford			59-B		29			
42	1945	Ford			59-B		29			

ROSTER NOTES

1. Cars 1-10 were given Indian names. Among them were 1-Wyoming; 2-Nay Aug; 3-Nescopeck; 6-Onas; 8-Oneida and 10-Capoose. Original car 10 was destroyed by fire in 1906; it was replaced with a new type M body made by the John Stephenson Car Co., Elizabeth, N.J. In 1907, type L cars were rebuilt with double-end Westinghouse unit switch control, including 251-C switch group. Cars 2 and 4 were sold to the Galveston-Houston Electric Railway on June 24, 1920, where they were renumbered 123 and 125 respectively. By 1926, only cars 8 and 10(II) remained on the roster; car 10(II) was retired in 1936 and scrapped in 1938.
2. Beginning in 1919, Westinghouse type 557-A motors and Westinghouse 251-C switch group replaced the original apparatus. Car 11 was wrecked in 1920; 12: body retired in 1931, trucks and control transferred to 38; 13: body retired in 1931, trucks and control transferred to 39; 14 and 15: retired 1926, trucks and control transferred to steel cars; 16: trucks and control transferred to 37 in 1929: 17 and 18: retired in 1935, one set of motors and trucks transferred to locomotive 402 in 1936, the other set retained as spares; 19 and 23: converted for one-man operation in 1938 for use on the Dunmore branch, listed as obsolete in 1954 and scrapped in 1955; 20 and 21: retired in 1926; 22: retired by 1942.
3. As noted above, the new steel car bodies were equipped with trucks and controls transferred from type M cars. All the steel cars, coaches 31-39 and combines 114-116, were sold to John C. Bauman of Childs, Pa. in 1953 for $12,600. They were scrapped in 1956.
4. Coaches 37-39 and combined 116 were constructed with one additional window panel, making them 2'-8" longer than cars 31-36 and 114-115.
5. Cars 37-39 used old trucks equipped with Hyatt roller bearings, which were gradually replaced with friction bearings.
6. Cars 101-104 were originally numbered 11-14. They were renumbered shortly after delivery. Original car 14 was diverted by the builder to the Grand Rapids, Grand Haven and Muskegon Railway, the only other contemporary third-rail Westinghouse project; another car was constructed for the L&WV in 1903. Car 101 was converted to a baggage trailer by 1927 and scrapped by 1938; 102: retired in 1925, body sold to G.T. Abel for $1,200, resold to Southern New York Power and Railway Co., Oneonta, N.Y.; 103: sold to Galveston-Houston Electric Railway on June 24, 1920 and renumbered 151; 104: retired by 1925, body burned in 1939.
7. Cars 101-104 and 111-113 were given names, usually Indian. Among them were 101-Luzerne; 103-Wampum; 104-Westmoreland and 111-Powhatan.
8. Car 113 damaged by fire in 1920; rebuilt by Jackson and Sharp at a cost of $6,800. Cars 111 and 113 received new trucks in 1926. Car 111 retired 1939; 112 retired in 1931; 113 retired by 1939.
9. Cars 114-116 used old trucks. 116 originally equipped with Hyatt roller bearings, which were gradually replaced by friction bearings. All combines had 10 additional seats in the baggage compartment.
10. Express cars were used on occasion to haul up to three loaded freight cars. Cars 201 and 203 were off the company's roster by 1926; both were probably sold to the Southern New York Power and Railway Co., Oneonta, N.Y. in 1925. 202 last operated on April 14, 1948, when express service was discontinued. 204 appeared on the roster in 1930 and was later used for parts; the body hulk survived as a storage shanty into the mid-1950's.
11. The L&WV called this hybrid car a work locomotive on its official equipment list. In 1905 it was equipped with a Priest flanger for snow removal. Occasionally, it was used to haul up to three freight cars. Wrecked in 1918 but rebuilt. Car still on roster in 1954 but listed in "bad condition." It was retired soon thereafter.
12. Car 302 was rebuilt from coal car 702 prior to 1926; it was used as a flat car and dismantled in 1935.
13. Car 303 was apparently rebuilt from a 700 series car prior to 1926; it was used as a weed-spraying car, with retirement date unknown.
14. Car 802 converted from a Rogers ballast car. Retired in 1954.
15. Locomotive 401 was the original joint venture of the Baldwin and Westinghouse companies in manufacturing electric locomotives. As constructed, it was an AC-DC machine. Rebuilt in 1906 and sold to L&WV. Out of service, August 19, 1953; scrapped in 1956.
16. Locomotive 402 was wrecked in 1918 and further damaged by fire in 1920. Trucks and electrical equipment were transferred to new locomotive 403 after the fire. 402 rebuilt in 1921 with passenger car motors and trucks. In 1931, it was equipped with motors from coaches 17 and 18. Out of service, August 19, 1953; scrapped in 1956.
17. As constructed in L&WV shops, locomotive 403 was to receive new trucks and old passenger motors. Instead, trucks and motors of fire-damaged 402 transferred to 403 in 1920. Out of service, August 19, 1953; scrapped in 1956.
18. Locomotive 4 was originally Lackawanna and Bloomsburg 4; renumbered as DL&W 204 in 1886. Sold in 1901 to George F. Lee Construction Co.; renumbered 4 and named "John R. Lee, Contractor." Removed from property after construction was completed. Lee Construction Co. occasionally used an additional locomotive numbered 5; no information is available on this machine.
19. Locomotive 10 was originally Morris and Essex 102; renumbered by DL&W in 1899 as 244. Purchased through Westinghouse, Church Kerr in 1902 and renumbered 10. Was used only on rare occasions after line opened. Sold to Luria Brothers of Reading, Pa. for scrap for $500 in 1911.
20. By 1928, car 502 converted for service as a line car. 504 was converted into a flat car around 1920 and renumbered as 1001(II). 507 wrecked in 1909; replaced by a duplicate car numbered 507(II) at a cost of $1,000. Trucks from the original car were used in the car house. 507(II) used as a work car by the roadway department through 1954. 500, 501, 503, 506, 508 and 509 were used through the early 1930's.
21. 601-605 were purchased to replace 1001-1005 and remained in service until 1954. 602 was later used as an idler for the derrick, 1919-1954. 605 was equipped with two tanks for use as a weed killer.
22. 1001 was destroyed in 1917 and replaced by 1001(II), ex-504. Trucks from 1001 were used in the car house. 1002 was used by the roadway department.
23. By 1917, car 1003 was serving as an ash car and 1004 was used by the electric line department. Disposition unknown. 1005 was listed as a work car in 1928, a derrick idler in 1929 and a flat car by 1954.
24. These cars were purchased used from the DL&W. 702 was rebuilt as work car 302 prior to 1926. 701, 703-710 scrapped in 1921; parts used to make ash cars 711-716(II).
25. Purchased used from the DL&W. Scrapped by 1921. Parts used to make ash cars 711-716(II).
26. Purchased used from the Erie Railroad. By 1921, six cars were worn out and scrapped. 724 was destroyed by fire in 1921; all were scrapped by 1930.
27. These hoppers were used in interchange service. 752, 754, 762 and 764 were still on the roster in 1954 but noted as obsolete; the others were scrapped earlier.
28. 711-716(II) were rebuilt from parts of original 701, 703-720. 716 scrapped by 1928, with the others gradually following. No information on 717-720(II).
29. In 1926, lead and motor trucks were renewed on 15 cars: 10, 12, 13, 14, 16, 17, 18, 19, 20, 21, 22, 23, 111, 112, 113. Some motor trucks and electrical control equipment transferred to steel cars 31-39 and 114-116.
30. All old wooden and steel equipment was disposed of in 1955 after court authorization. Included were 8 ash cars, 4 hopper cars, 2 wood passenger cars (19 & 23), 5 low side cars, 2 flat cars and 2 miscellaneous cars. Scrap metal totaled 512,200 lbs.
31. Motor bus records are scattered and incomplete. The roster reflects available information.

Bibliography

ARCHIVES AND PRIMARY SOURCES

Lackawanna and Wyoming Valley Railroad. Records and Papers. George Arents Research Library, Syracuse Univesity.

Lackawanna and Wyoming Valley Railroad. Edward S. Miller Collection.

McGraw Electric Railway Directory, 1924. New York: McGraw-Hill Co., 1924.

Official Guide of the Railways and Steam Navigation Lines of the United States. Various issues. New York: National Railway Publication Company, 1910-1958.

Pennsylvania. Bureau of Statistics. *Pennsylvania Statistical Abstract,* 16th Edition. Harrisburg: Bureau of Statistics, Research and Planning, 1974.

Pennsylvania. Department of Environmental Resources. *Annual Report on Mining, Oil and Gas and Land Reclamation and Conservation Activities.* Harrisburg: Department of Environmental Resources, 1979.

Pennsylvania. Department of General Services. *The Pennsylvania Manual,* Vol. 98. Harrisburg: Department of General Services, 1967.

SECONDARY SOURCES

Archer, Robert F. *The History of the Lehigh Valley Railroad.* Berkeley, Calif.: Howell-North Books, 1977.

Baldwin-Westinghouse Electric Locomotives, 1925. Chicago: Electric Railway Historical Society, 1957.

Bradsley, H.C., editor. *History of Luzerne County, Pa.* Chicago: S.B. Nelson & Co., 1897.

Burgess, George H. and Kennedy, Miles C. *Centennial History of the Pennsylvania Railroad Company, 1846-1946.* Philadelphia: The Pennsylvania Railroad Company, 1949.

Carlson, Norman and Peterson, Arthur, editors. *Remember When Trolley Wires Spanned the Country.* Chicago: Central Electric Railfans Assn., 1980.

Condit, Carl W. *The Port of New York.* Chicago: University of Chicago Press, 1980.

Craft, David. *History of Scranton, Pa.* Dayton: United Brethren Publishing House, 1981.

Delaware and Hudson Company. *A Century of Progress: A History of the Delaware and Hudson Company, 1823-1923.* Albany: J.B. Lyon Co., 1925.

Ellis, William and Miller, Edward S., "The Laurel Line," *Era Headlights.* December 1949.

Ford, Bacon and Davis. *For Human Needs.* New York: Privately printed, 1967.

Helmer, William F. *O & W: The Long Life and Slow Death of the New York, Ontario and Western Railway.* Berkeley: Howell-North Books, 1959.

Hilton, George W. and Due, John F. *The Electric Interurban Railway in America.* Stanford: Stanford University Press, 1960.

Hudson Coal Company. *The Story of Anthracite.* New York: The Hudson Coal Co., 1932.

Hungerford, Edward. *Men of Erie.* New York: Random House, 1946.

Kulp, Randolph L., editor. *Railroads in the Lehigh River Valley,* revised edition. Allentown: Lehigh Valley Chapter, NRHS, 1962.

Lucas, Walter P. *The History of the New York, Susquehanna and Western Railroad,* second edition. Roselle Park: Railroadians of America, 1939.

Middleton, William D. *The Interurban Era.* Milwaukee: Kalmbach Publishing Co., 1961.

________. *The Time of the Trolley.* Milwaukee: Kalmbach Publishing Co., 1967.

Munsell, W.W. Co. *History of Luzerne, Lackawanna and Wyoming Counties, Pa.* New York: W.W. Munsell & Co., 1880.

Murphy, Thomas. *Lackawanna County.* Indianapolis: Historical Publishing Co., 1928.

Parkes, Henry B. and Carosso, Vincent P. *Recent America, A History.* 2 volumes. New York: Thomas Y. Crownell Co., 1963.

Prout, Henry G. *A Life of George Westinghouse.* New York: The Scribner Press, 1921.

Railroadians of America. *New York, Susquehanna and Western and Wilkes-Barre and Eastern: The Next Station Will Be.* Vol. 5. Roselle Park: Railroadians of America, Inc., 1977.

Rohrback, Benson W. *Pennsylvania's Street Railways.* West Chester: Ben Rohrbeck Traction Publications, 1982.

Saylor, Roger B. *The Railroads of Pennsylvania.* Bureau of Business Research, College of Business Administration. University Park: The Pennsylvania University Press, 1964.

Shaughnessy, Jim. *Delaware and Hudson.* Berkeley: Howell-North Books, 1967.

Taber, Thomas T. *The Delaware, Lackawanna and Western Railroad in the Nineteenth Century.* Muncy: Published by the author, 1977.

Taber, Thomas T. and Taber, Thomas T. III. *The Delaware, Lackawanna and Western Railroad in the Twentieth Century, 1899-1960.* 2 volumes. Muncy: Published by the author, 1981.

Throop, Benjamin H. *A Half Century in Scranton.* Scranton: Scranton Republican 1895.

Wakefield, Manville B. *Coal Boats to Tidewater.* South Fallsburg, N.Y.: Steingart Associates, Inc., 1965.

Westinghouse Electric and Manufacturing Company. "Results of Electrification," Special Publication 1505 (April 1951). Reprinted in *Westinghouse Electric Railway Transportation.* Chicago: Central Electric Railfans Association, 1979.

Westinghouse Electric Corporation. *George Westinghouse, 1846-1914.* East Pittsburgh: Westinghouse Electric Corporation, 1946.

Wish, Harvey. *Contemporary America.* New York: Harper and Row, 1958.

Woods, Herb. *Galveston-Houston Electric Railway.* Glendale: Interurban Press, 1959.

PERIODICALS AND NEWSPAPERS
Electric Railway Journal, 1926.
Headlights, various issues, 1945-1955.
New York Times, 1927, 1928, 1933, 1947, 1957.
Philadelphia Inquirer, 1962.
Pittston Daily Gazette, various issues, 1931-1953.
Pittston Independent, various issues, 1949.
Pittston Sunday Dispatch, various issues, 1953-1972.
Railway Age, 1918.
Street Railway Journal, various issues, 1903-1908.
The Scrantonian, 1914.
Scranton Republican, various issues, 1903-1931.
Scranton Times, various issues, 1901-1961.
Scranton Tribune, various issues, 1903-1957.
Wilkes-Barre Evening News, various issues, 1913-1946.
Wilkes-Barre Independent, 1951.
Wilkes-Barre Record, various issues, 1903-1966.
Wilkes-Barre Sunday News, 1903.
Wilkes-Barre Times, 1907.
Wilkes-Barre Times-Leader, various issues, 1914-1977.

INTERVIEWS
Edgar S. Boland
E. McLain Watters, Jr.

Notes on Sources

The major source for this work is the extensive collection of the Lackawanna and Wyoming Valley Railroad Company records and papers, held by the George Arents Research Library, Syracuse University. While weak in a few areas, this collection documents most of the company's activities during its more than half century of independent existence. A much smaller but still significant collection of company papers is held by Edward S. Miller, which was also used extensively.

Statistics used in the financial summaries represent composite figures from a number of different sources within the company records. In some instances, they are inconsistent and totals are not accurate because the reports contained variations and the authors rounded figures to simplify the tables, which are intended only to provide a general picture of the company's financial status within revenue and cost categories.

Chapter I

Sources include the various railroad and county histories listed in the bibliography, Scranton and Wilkes-Barre newspapers, the *Street Railway Journal, Headlights* and the L&WV Collection, Syracuse University.

Chapter II

This chapter is based primarily on numerous letters, reports, memoranda and other data in the L&WV Collection, Syracuse University. Additional information was gained from Scranton and Wilkes-Barre newspapers, the *Street Railway Journal,* Condit's *Port of New York* and *Moody's Manual of Investments; Public Utilities, 1926.*

Chapter III

Most information in this chapter is based on the letters, records, reports and agreements in the L&WV Collection. A few details were found in the Scranton and Wilkes-Barre newspapers, *Headlights* and Wish, *Contemporary America.*

Chapter IV

The L&WV Collection's reports, letters and memos are the primary sources. Also useful were the Scranton newspapers, *New York Times, Headlights* and *Electric Railway Journal.*

Chapter V

The company files in the L&WV Collection provided most of the information. Also useful were the Scranton and Wilkes-Barre newspapers, *New York Times* and the Miller Collection.

Chapter VI

This chapter is based on information from the company records in the L&WV Collection and the Miller Collection. Additional facts were taken from the Scranton, Wilkes-Barre, Pittston and Philadelphia newspapers; *Headlights* and Commonwealth of Pennsylvania reports. E. McLain Watters, Jr., provided further details in an interview on December 27, 1982.

Chapter VII

Some information was gained from the L&WV Collection; also useful were reports in local newspapers, the *New York Times, Headlights* and the Miller Collection.

Index

LACKAWANNA & WYOMING VALLEY RAILROAD COMPANY 114 AND 115

SCALE: 3/8″ = 1′-0″

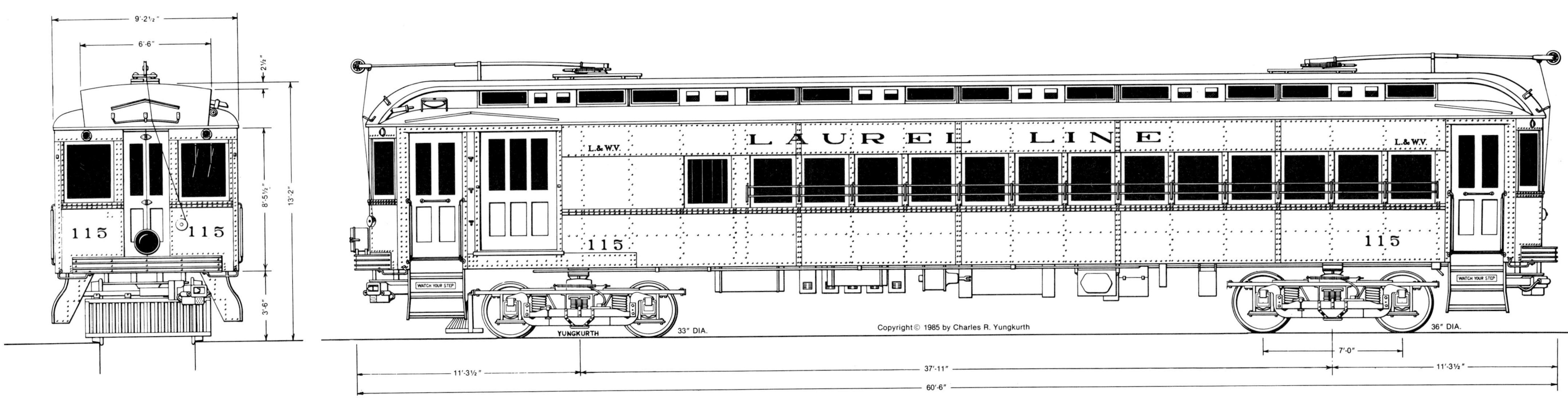

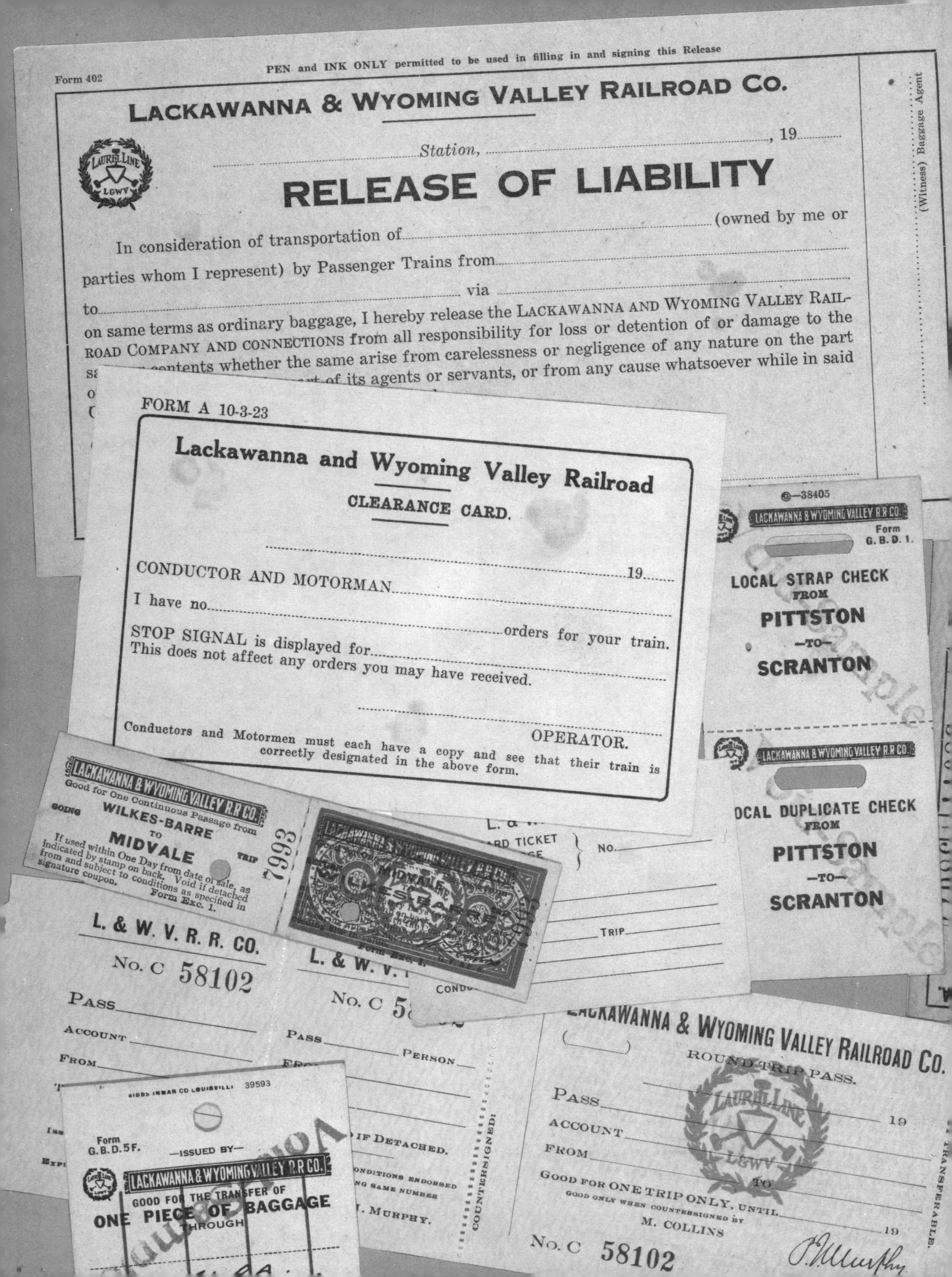

Form 402
PEN and INK ONLY permitted to be used in filling in and signing this Release
LACKAWANNA & WYOMING VALLEY RAILROAD CO.
Station, , 19
RELEASE OF LIABILITY
In consideration of transportation of (owned by me or parties whom I represent) by Passenger Trains from
to via
on same terms as ordinary baggage, I hereby release the LACKAWANNA AND WYOMING VALLEY RAILROAD COMPANY AND CONNECTIONS from all responsibility for loss or detention of or damage to the
contents whether the same arise from carelessness or negligence of any nature on the part
of its agents or servants, or from any cause whatsoever while in said
(Witness) Baggage Agent
FORM A 10-3-23
Lackawanna and Wyoming Valley Railroad
CLEARANCE CARD.
19
CONDUCTOR AND MOTORMAN
I have no
orders for your train.
STOP SIGNAL is displayed for
This does not affect any orders you may have received.
OPERATOR.
Conductors and Motormen must each have a copy and see that their train is correctly designated in the above form.
6—38405
LACKAWANNA & WYOMING VALLEY R.R.CO.
Form G.B.D.1.
LOCAL STRAP CHECK
FROM
PITTSTON
—TO—
SCRANTON
LACKAWANNA & WYOMING VALLEY R.R.CO.
LOCAL DUPLICATE CHECK
FROM
PITTSTON
—TO—
SCRANTON
LACKAWANNA & WYOMING VALLEY R.R.CO.
Good for One Continuous Passage from
GOING
WILKES-BARRE
TO
MIDVALE
TRIP
If used within One Day from date of sale, as indicated by stamp on back. Void if detached from and subject to conditions as specified in signature coupon.
Form Exc. 1.
7993
L. & W. V. R. R. CO.
No. C 58102
PASS
ACCOUNT
FROM
NO.
TRIP
IF DETACHED.
M. MURPHY.
COUNTERSIGNED:
LACKAWANNA & WYOMING VALLEY RAILROAD CO.
ROUND TRIP PASS.
PASS 19
ACCOUNT
FROM
TO
GOOD FOR ONE TRIP ONLY, UNTIL 19
GOOD ONLY WHEN COUNTERSIGNED BY
M. COLLINS
No. C 58102
NON-TRANSFERABLE.
39593
Form G.B.D.5F.
—ISSUED BY—
LACKAWANNA & WYOMING VALLEY R.R.CO.
GOOD FOR THE TRANSFER OF
ONE PIECE OF BAGGAGE
THROUGH